Brickwork

NVQ and Technical Certificate Level 2

www.heinemann.co.uk
✓ Free online support
✓ Useful weblinks
✓ 24 hour online ordering

01865 888058

Heinemann, Halley Court, Jordan Hill, Oxford OX2 8EJ

Heinemann is the registered trademark of Harcourt Ltd

© Carillion Construction Ltd

First published 2006

11 10 09 08 07 06

10 9 8 7 6 5 4 3 2 1

British Library Cataloguing in Publication Data is available from the British Library on request.

10-digit ISBN: 0 435 43086 6

13-digit ISBN: 978 0 435430 86 3

Copyright notice

All rights reserved. No part of this publication may be reproduced in any form or by any means (including photocopying or storing it in any medium by electronic means and whether or not transiently or incidentally to some other use of this publication) without the written permission of the copyright owner, except in accordance with the provisions of the Copyright, Designs and Patents Act 1988 or under the terms of a licence issued by the Copyright Licensing Agency, 90 Tottenham Court Road, London W1T 4LP. Applications for the copyright owner's written permission should be addressed to the publisher.

Designed by HL Studios
Layout by HL Studios
Printed in the UK by Scotprint
Illustrated by HL Studios

Cover design by GD Associates
Cover photo: © Harcourt Ltd/Gareth Boden and Corbis

Websites

Please note that the examples of websites suggested in this book were up-to-date at the time of writing. We have made all links available on the Heinemann website at www.heinemann.co.uk/hotlinks. When you access the site, the express code is 0866P.

The information and activities in this book have been prepared according to the standards reasonably to be expected of a competent trainer in the relevant subject matter. However, you should be aware that errors and omissions can be made and that different employers may adopt different standards and practices over time. Therefore, before doing any practical activity, you should always carry out your own Risk Assessment and make your own enquiries and investigations into appropriate standards and practices to be observed.

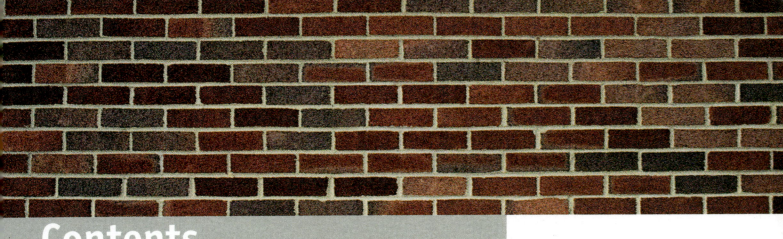

Contents

Acknowledgements — iv
Introduction — v

Chapter 1   The construction industry — 1
Chapter 2   Health and safety — 33
Chapter 3   Working at height — 67
Chapter 4   Drawings — 85
Chapter 5   Hand tools and equipment — 103
Chapter 6   Handling and storage of materials — 121
Chapter 7   Setting out — 141
Chapter 8   Mixing mortar — 159
Chapter 9   Bonding — 169
Chapter 10   Laying bricks and blocks to a line — 179
Chapter 11   Solid walls — 191
Chapter 12   Cavity walling — 207
Chapter 13   Arch construction — 229
Chapter 14   Pointing and jointing — 247
Chapter 15   Drainage systems — 257
Chapter 16   Basic concreting — 279

Glossary — 303
Index — 305

Acknowledgements

Carillion would like to thank the following people for their contribution to this book: David Whitten, Kevin Jarvis and Ralph Need.

Harcourt would like to thank everyone at the Carillion Construction Training Centre in Sunderland for all their help at the photo shoots.

Special thanks to Chris Ledson at Toolbank for supplying some photos. Visit the Toolbank website at www.toolbank.com.

Photos

Alamy Images p252; Construction photography p38, p117, p291 (top), p299 (top); Construction photography/Dave Potter p72 (bottom); Corbis p1, p33, p57, p59 (top), p67, p71, p85, p103, p141, p159, p198, p199, p245 (bottom); Getty Images/Photodisc p7, p10, p14, p16, p17, p28, p42, p43, p70, p169, p229; Harcourt Ltd/Chris Honeywell p58; Harcourt Ltd/Ginny Stroud-Lewis p60, p61 (top); IStockPhoto/Guy Erwood p72 (top); Maria Joannou p165, p285 (bottom), p291 (bottom); Photographers Direct/David Griffiths p297 (top); Photographers Direct/Roger G. Howard p207, p285; Photographers Direct/Robert Kawka p245 (top); Photos.com p244, p299 (bottom); Shout, p52 (top); Toolbank p59 (bottom), p61 (bottom); David Whitten p149; Rachael Williams p2.

All other photos copyright Gareth Boden/Harcourt Ltd.

About this book

This book has been written based on a concept used within Carillion Training Centres for many years. That concept is about providing learners with the necessary information they need to support their studies and at the same time ensuring it is presented in a style which they find both manageable and relevant.

The content of this book has been put together by a team of instructors, each of whom have a wealth of knowledge and experience in both training for NVQs and Technical Certificates and their trade.

This book has been produced to help the learner build a sound knowledge and understanding of all aspects of the NVQ and Technical Certificate requirements associated with their trade. It has also been designed to provide assistance when revising for Technical Certificate end tests and NVQ job knowledge tests.

Each chapter of this book relates closely to a particular unit of the NVQ or Technical Certificate and aims to provide just the right level of information needed to form the required knowledge and understanding of that subject area.

This book provides a basic introduction to the tools, materials and methods of work required to enable you to complete work activities effectively and productively. Upon completion of your studies, this book will remain a valuable source of information and support when carrying out your work activities.

For further information on how the content of this student book matches to the unit requirements of the NVQ and Intermediate Construction Award, please visit www.heinemann.co.uk and follow the FE and Vocational link, followed by the Construction link, where a detailed mapping document is available for download.

How this book can help you

You will discover a variety of features throughout this book, each of which have been designed and written to increase and improve your knowledge and understanding. These features are:

- **Photographs** – many photographs that appear in this book are specially taken and will help you to follow a step-by-step procedure or identify a tool or material.

- **Illustrations** – clear and colourful drawings will give you more information about a concept or procedure.

- **Definitions** – new or difficult words are picked out in **bold** in the text and defined in the margin.

- **Remember** – key concepts or facts are highlighted in these margin boxes.

- **Find out** – carry out these short activities and gain further information and understanding of a topic area.

- **Did you know?** – interesting facts about the building trade.

- **Safety tips** – follow the guidance in these margin boxes to help you work safely.

- **FAQs** – frequently asked questions appear in all chapters along with informative answers from the experts.

- **On the job scenarios** – read about a real-life situation and answer the questions at the end. What would you do? (Answers can be found in the Tutor Resource Disc that accompanies this book.)

- **End of chapter knowledge checks** – test your understanding and recall of a topic by completing these questions.

- **Glossary** – at the end of this book you will find a comprehensive glossary that defines all the **bold** words and phrases found in the text. A great quick reference tool.

- **Links to useful websites** – any websites referred to in this book can be found at www.heinemann.co.uk/hotlinks. Just enter the express code

chapter 1

The construction industry

OVERVIEW

Construction means creating buildings and services. These might be houses, hospitals, schools, offices, roads, bridges, museums, prisons, train stations, airports, monuments – and anything else you can think of that needs designing and building! What about an Olympic stadium? The 2012 London games will bring a wealth of construction opportunity to the UK and so it is an exciting time to be getting involved.

In the UK, 2.2 million people work in the construction industry – more than in any other – and it is constantly expanding and developing. There are more choices and opportunities than ever before and pay and conditions are improving all the time. Your career doesn't have to end in the UK either – what about taking the skills and experience you are developing abroad? Construction is a career you can take with you wherever you go. There's always going to be something that needs building!

This chapter will cover the following:

- Understanding the industry
- Communication
- Getting involved in the construction industry
- Sources of information and advice.

Understanding the industry

The construction industry is made up of countless companies and businesses that all provide different services and materials. An easy way to divide these companies into categories is according to their size.

- A small company is defined as having between 1 and 49 members of staff.
- A medium company consists of between 50 and 249 members of staff.
- A large company has 250 or more people working for it.

A business might only consist of one member of staff (a sole trader).

Find out

Think of an example of a small, medium and large construction company. Do you know of any construction companies that have only one member of staff?

The different types of construction work

There are four main types of construction work:

1. **New work** – this refers to a building that is about to be or has just been built.

2. **Maintenance work** – this is when an existing building is kept up to an acceptable standard by fixing anything that is damaged so that it does not fall into disrepair.

3. **Refurbishment/renovation work** – this generally refers to an existing building that has fallen into a state of disrepair and is then brought up to standard by repair. It also refers to an existing building that is to be used for a different purpose, for example, changing an old bank into a pub.

4. **Restoration work** – this refers to an existing building that has fallen into a state of disrepair and is then brought back to its original condition or use.

New work is just one type of construction area

These four types of work can fall into one of two categories depending upon who is paying for the work:

1. Public – the government pays for the work, as is the case with most schools and hospitals etc.

2. Private – work is paid for by a private client and can range from extensions on existing houses to new houses or buildings.

Job and careers

Jobs and careers in the construction industry fall mainly into one of four categories:

1. building

2. civil engineering

3. electrical engineering

4. mechanical engineering.

Building involves the physical construction (making) of a structure. It also involves the maintenance, restoration and refurbishment of structures.

Civil engineering involves the construction and maintenance of work such as roads, railways, bridges etc.

Electrical engineering involves the installation and maintenance of electrical systems and devices such as lights, power sockets and electrical appliances etc.

Mechanical engineering involves the installation and maintenance of things such as heating, ventilation and lifts.

The category that is the most relevant to your course is building.

Brickwork NVQ and Technical Certificate Level 2

Bricklayers are building craft workers

Job types

The construction industry employs people in four specific areas:

1. professionals
2. technicians
3. building craft workers
4. building operatives.

Professionals

Professionals are generally of graduate level (i.e. people who have a degree from a university) and may have one of the following types of job in the construction industry:

- architect – someone who designs and draws the building or structure
- structural engineer – someone who oversees the strength and structure of the building
- surveyor – someone who checks the land for suitability to build on
- service engineer – someone who plans the services needed within the building, for example, gas, electricity and water supplies.

Technicians

Technicians link professional workers with craft workers and are made up of the following people:

- architectural technician – someone who looks at the architect's information and makes drawings that can be used by the builder
- building technician – someone who is responsible for estimating the cost of the work and materials and general site management
- quantity surveyor – someone who calculates ongoing costs and payment for work done.

Building craft workers

Building craft workers are the skilled people who work with materials to physically construct the building. The following jobs fall into this category:

- carpenter or joiner – someone who works with wood but also other construction materials such as plastic and iron. A carpenter primarily works on site while a joiner usually works off site, producing components such as windows, stairs, doors, kitchens, and **trusses**, which the carpenter then fits into the building
- bricklayer – someone who works with bricks, blocks and cement to build the structure of the building
- plasterer – someone who adds finish to the internal walls and ceilings by applying a **plaster skim**. They also make and fix plaster **covings** and plaster decorations
- painter and decorator – someone who uses paint and paper to decorate the internal plaster and timberwork such as walls, ceilings, windows and doors, as well as **architraves** and **skirting**
- electrician – someone who fits all electrical systems and fittings within a building, including power supplies, lights and power sockets
- plumber – someone who fits all water services within a building, including sinks, boilers, water tanks, radiators, toilets and baths. The plumber also deals with lead work and rainwater fittings such as guttering
- slater and tiler – someone who fits tiles on to the roof of a building, ensuring that the building is watertight
- woodworking machinist – someone who works in a machine shop, converting timber into joinery components such as window sections, spindles for stairs, architraves and skirting boards, amongst other things. They use a variety of machines such as lathes, bench saws, planers and sanders.

Definition

Trusses – prefabricated components of a roof which spread the load of a roof over the outer walls and forms its shape

Plaster skim – a thin layer of plaster that is put on to walls to give a smooth and even finish

Covings – a decorative moulding that is fitted at the top of a wall where it meets the ceiling

Architraves – a decorative moulding, usually made from timber, that is fitted around door and window frames to hide the gap between the frame and the wall

Skirting – a decorative moulding that is fitted at the bottom of a wall to hide the gap between the wall and the floor

Building operatives

There are two different building operatives working on a construction site.

1. Specialist building operative – someone who carries out specialist operations such as dry wall lining, asphalting, scaffolding, floor and wall tiling and glazing.

2. General building operative – someone who carries out non-specialist operations such as kerb laying, concreting, path laying and drainage. These operatives also support other craft workers and do general labouring. They use a variety of hand tools and power tools as well as **plant**, such as dumper trucks and JCBs.

The building team

Constructing a building or structure is a huge task that needs to be done by a team of people who all need to work together towards the same goal. The team of people is often known as the building team and is made up of the following people.

Clients

The client is the person who requires the building or refurbishment. This person is the most important person in the building team because they finance the project fully and without the client there is no work. The client can be a single person or a large organisation.

Architect

The architect works closely with the client, interpreting their requirements to produce contract documents that enable the client's wishes to be realised.

Clerk of works

Selected by the architect or client to oversee the actual building process, the clerk of works ensures that construction sticks to agreed deadlines. They also monitor the quality of workmanship.

Definition

Plant – industrial machinery

Local Authority

The Local Authority is responsible for ensuring that construction projects meet relevant planning and building legislation. Planning and building control officers approve and inspect building work.

Quantity surveyor

The quantity surveyor works closely with the architect and client, acting as an accountant for the job. They are responsible for the ongoing evaluation of cost and interim payments from the client, establishing whether or not the contract is on budget. The quantity surveyor will prepare and sign off final accounts when the contract is complete.

The building team is made up of many different people

Specialist engineers

Specialist engineers assist the architect in specialist areas, such as civil engineering, structural engineering and service engineering.

Health and safety inspectors

Employed by the Health and Safety Executive (HSE), health and safety inspectors ensure that the building contractor fully implements and complies with government health and safety legislation. For more information on health and safety in the construction industry, see Chapter 2 (page 33).

Building contractors

The building contractors agree to carry out building work for the client. Contractors will employ the required workforce based on the size of the contract.

Estimator

The estimator works with the contractor on the cost of carrying out the building contract, listing each item in the bill of quantities (e.g. materials, labour and plant). They calculate the overall cost for the contractor to complete the contract, including further costs as overheads, such as site offices, management administration and pay, not forgetting profit.

Site agent

The site agent works for the building contractor and is responsible for the day-to-day running of the site such as organising deliveries etc.

Suppliers

The suppliers work with the contractor and estimator to arrange the materials that are needed on site and ensure that they are delivered on time and in good condition.

General foreman

The general foreman works for the site manager and is responsible for co-ordinating the work of the ganger (see below), craft foreman and subcontractors. They may also be responsible for the hiring and firing of site operatives. The general foreman also liaises with the clerk of works.

Craft foreman

The craft foreman works for the general foreman organising and supervising the work of particular crafts. For example, the carpentry craft foreman will be responsible for all carpenters on site.

Ganger

The ganger supervises the general building operatives.

Chargehand

The chargehand is normally employed only on large build projects, being responsible for various craftsmen and working with joiners, bricklayers, and plasterers.

Operatives

Operatives are the workers who carry out the building work, and are divided into three subsections:

1. Craft operatives are skilled tradesman such as joiners, plasterers, bricklayers.

2. Building operatives include general building operatives who are responsible for drain laying, mixing concrete, unloading materials and keeping the site clean.

3. Specialist operatives include tilers, pavers, glaziers, scaffolders and plant operators.

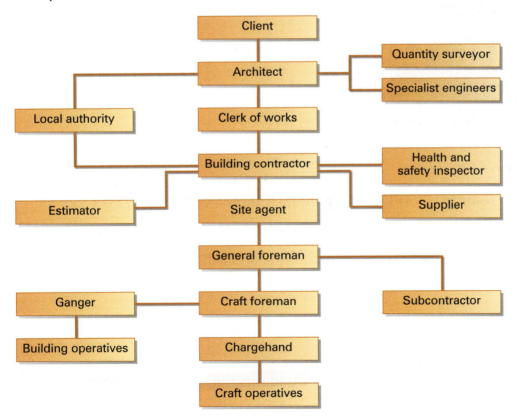

Figure 1.1 The building team

The different types of building

There are of course lots of very different types of building, but the main types are:

- residential – houses and flats etc.
- commercial – shops and supermarkets etc.
- industrial – warehouses and factories etc.

A low rise residential building

These types of building can be further broken down by the height or number of storeys that they have (one storey being the level from floor to ceiling):

- low rise – a building with one to three storeys
- medium rise – a building with four to seven storeys
- high rise – a building with seven storeys or more.

Buildings can also be categorised according to the number of other buildings they are attached to:

- detached – a building that stands alone and is not connected to any other building
- semi-detached – a building that is joined to one other building and shares a dividing wall, called a party wall
- terraced – a row of three or more buildings that are joined together, of which the inner buildings share two party walls.

Building requirements

Every building must meet the minimum requirements of the *Building Regulations*, which were first introduced in 1961 and then updated in 1985. The purpose of building regulations is to ensure that safe and healthy buildings are constructed for the public and that **conservation** is taken into account when they are being constructed. Building regulations enforce a minimum standard of building work and ensure that the materials used are of a good standard and fit for purpose.

> **Definition**
>
> **Conservation** – preservation of the environment and wildlife

What makes a good building?

When a building is designed, there are certain things that need to be taken into consideration, such as:

- security
- safety
- privacy
- warmth
- light
- ventilation.

A well designed building will meet the minimum standards for all of the considerations above and will also be built in line with building regulations.

Identifying the different parts of a building

All buildings consist of the following two main parts:

1. the substructure
2. the superstructure.

The substructure consists of all building work below the ground level, including the foundations, up to the **damp proof course**. The purpose of the substructure is to spread the load of the building.

The superstructure consists of all the building work above the substructure and its purpose is to provide shelter and divide space.

The things that make up the substructure or superstructure can be divided into four different sections:

1. Primary elements – these include the main parts of the building that provide support, protection, floor-to-floor access and the division of space. Examples of primary elements are foundations, walls, floors, roofs and stairs.

2. Secondary elements – these include the non-essential and non-load bearing parts that are used to close off openings or to provide a finish. Examples of secondary elements are doors, windows, skirting and architraves.

3. Finishing elements – these include the final parts required to complete a component and can be superficial or necessary to complete the job. Examples of finishing elements are paint, wallpaper, plaster or face brickwork.

4. Services – these are the electrical, mechanical and specialist installations that are normally piped or wired into the building. Examples of services are running water and electricity.

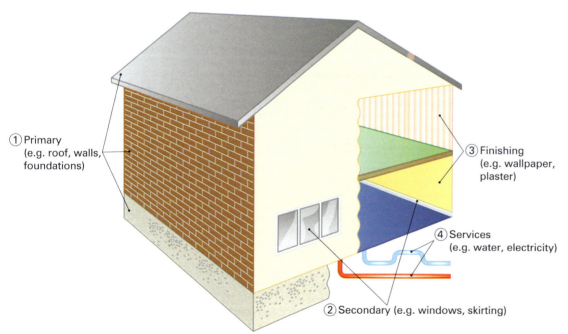

Figure 1.2 The four elements of a building

Communication

Communication, in the simplest of terms, is a way or means of passing on information from one person to another. Communication is very important in all areas of life and we often do it without even thinking about it. You will need to communicate well when you are at work, no matter what job you do. What would happen if someone couldn't understand something you had written or said? If we don't communicate well, how will other people know what we want or need and how will we know what other people want?

Companies that do not establish good methods of communicating with their workforce or with other companies, will not function properly and will end up with bad working relationships. Good working relationships can *only* be achieved with co-operation and good communication.

Methods of communication

There are many different ways of communicating with others and they all generally fit into one of these three categories:

1. speaking (verbal communication), for example talking face to face or over the telephone

2. writing, for example sending a letter or taking a message

3. body language, for example the way we stand or our facial expressions.

Each method of communicating has good points (advantages) and bad points (disadvantages).

Verbal communication

Verbal communication is the most common method we use to communicate with each other. If two people don't speak the same language or if someone speaks very quietly or not very clearly, verbal communication cannot be effective. Working in the construction industry you may communicate verbally with other people face to face, over the telephone or by radio/walkie-talkie.

Verbal communication is probably the method you will use most

Advantages

Verbal communication is instant, easy and can be repeated or rephrased until the message is understood.

Disadvantages

Verbal communication can be easily forgotten as there is no physical evidence of the message. Because of this it can be easily changed if passed to other people. Someone's accent or use of slang language can sometimes make it difficult to understand what they are saying.

Written communication

Written communication can take the form of letters, faxes, messages, notes, instruction leaflets, text messages, faxes, drawings and emails, amongst others.

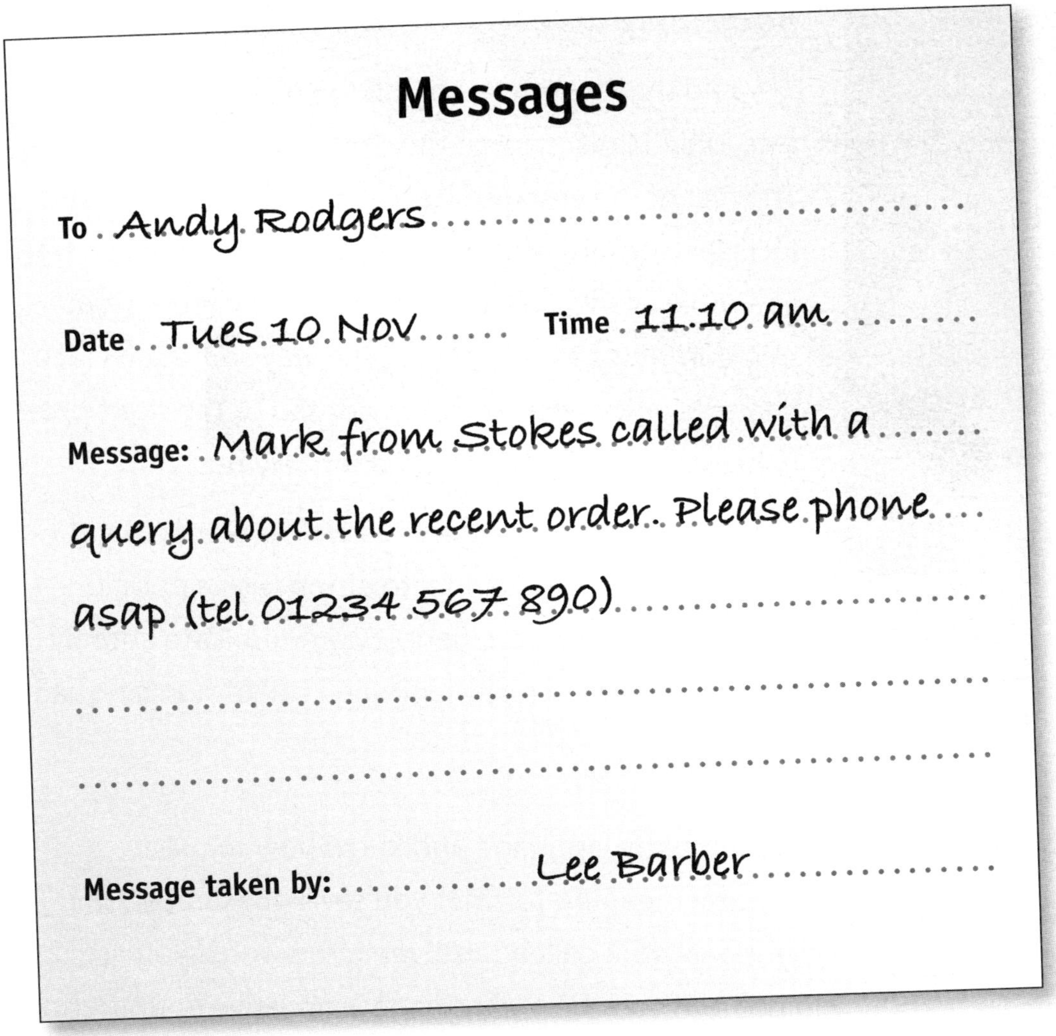

Chapter 1 The construction industry

> # Messages
>
> To Andy Rodgers.............................
>
> Date ..Tues.10.Nov...... Time .11.10.am.........
>
> Message: ..Mark.from.Stokes.called.with.a.......
> query.about.the.recent.order..Please.phone....
> asap.(tel.01234.567.890).....................
> ....
> ....
>
> Message taken by: .... Lee Barber.............

Figure 1.3 A message is a form of written communication

Advantages

There is physical evidence of the communication and the message can be passed on to another person without it being changed. It can also be read again if it is not understood.

Disadvantages

Written communication takes longer to arrive and understand than verbal communication and body language. It can also be misunderstood or lost. If it is handwritten, the reader may not be able to read the writing if it is messy.

Try to be aware of your body language

Body language

It is said that, when we are talking to someone face to face, only 10 per cent of the communication is verbal. The rest of the communication is body language and facial expression. This form of communication can be as simple as the shaking of a head from left to right to mean 'no' or as complex as the way someone's face changes when they are happy or sad or the signs given in body language when someone is lying.

We often use hand gestures as well as words to get across what we are saying, to emphasise a point or give a direction. Some people communicate entirely through a form of body language called sign language.

Advantages

If you are aware of your own body language and know how to use it effectively, you can add extra meaning to what you say. For example, say you are talking to a client or a work colleague. Even if the words you are using are friendly and polite, if your body language is negative or unfriendly, the message that you are giving out could be misunderstood. By simply maintaining eye contact, smiling and not folding your arms, you have made sure that the person you are communicating with has not got a mixed or confusing message.

Body language is quick and effective. A wave from a distance can pass on a greeting without being close, and using hand signals to direct a lorry or a load from a crane is instant and doesn't require any equipment such as radios.

Disadvantages

Some gestures can be misunderstood, especially if they are given from very far away, and gestures that have one meaning in one country or culture can have a completely different meaning in another.

Chapter 1 The construction industry

Which type of communication should I use?

Of the many different types of communication, the type you should use will depend upon the situation. If someone needs to be told something formally, then written communication is generally the best way. If the message is informal, then verbal communication is usually acceptable.

The way that you communicate will also be affected by who it is you are communicating with. You should of course always communicate in a polite and respectful manner with anyone you have contact with, but you need to also be aware of the need to sometimes alter the style of your communication. For example, when talking to a friend, it may be fine to talk in a very informal way and use slang language, but in a work situation with a client or a colleague, it is best to alter your communication to a more formal style in order to show professionalism. In the same way, it may be fine to leave a message or send a text to a friend that says 'C U @ 8 4 work', but if you wrote this down for a work colleague or a client to read, it would not look very professional and they may not understand it.

You will work with people from other trades

Communicating with other trades

Communicating with other trades is vital because they need to know what you are doing and when, and you need to know the same information from them. Poor communication can lead to delays and mistakes, which can both be costly. It is quite possible for poor communication to result in work having to be stopped or redone. Say you are decorating a room in a new building. You are just about to finish when you find out that the electrician, plumber and carpenter have to finish off some work in the room. This information didn't reach you and now the decorating will have to be done again once the other work has been finished. What a waste of time and money. A situation like this can be avoided with good communication between the trades.

Common methods of communicating in the construction industry

A career in construction means that you will often have to use written documents such as drawings, specifications and schedules. These documents can be very large and seem very complicated but, if you understand what they are used for and how they work, using such documents will soon become second nature.

For more detailed information see Chapter 4 Drawings page 85.

Drawings

Drawings are done by the architect and are used to pass on the client's wishes to the building contractor. Drawings are usually done to scale, whereby the drawing is drawn to a certain scale because it would be impossible to draw a full-sized version of the project. A common scale is 1:10, which means that a line 10 mm long on the drawing represents 100 mm in real life. Drawings often contain symbols instead of written words to get the maximum amount of information across without cluttering the page. See Chapter 4 pages 93 and 94 for more information.

Specifications

Specifications accompany a drawing and give you the sizes that are not available on the drawing, as well as telling you the type of material to be used and the quality that the work has to be finished to.

Schedules

A schedule is a list of repeated design information used on big building sites when there are several types of similar room or house. For example, a schedule will tell you what type of door must be used and where. Another form of schedule used on building sites contains a detailed list of dates by which work must be carried out and materials delivered etc.

Other documents

As well as drawings, specifications and schedules there are some other important types of documents you will come across that are not specifically about the building or structure you are working on. Rather, they are about your day-to-day tasks and your job. We will now look at a selection of these documents.

Timesheet

A timesheet is used to record the hours you have worked and where the work was carried out. Failure to complete your timesheet accurately and submit it on time may result in a loss of wages.

P. Gresford Building Contractors

Timesheet _____

Employee _____   **Project/site** _____

Date	Job no.	Start time	Finish time	Total time	Travel time	Expenses
M						
Tu						
W						
Th						
F						
Sa						
Su						
Totals						

Employee's signature _____

Supervisor's signature _____

Date _____

Figure 1.4 A typical timesheet

Brickwork NVQ and Technical Certificate Level 2

P. Gresford Building Contractors

Jobsheet

Customer   Chris MacFarlane

Address   1 High Street
Any Town
Any County

Work to be carried out

Repoint front of property

Special conditions/instructions

Check with client before spray-
treating if insect attack detected

Figure 1.5 A typical jobsheet

Jobsheet/Day worksheet

A jobsheet is used to record work to be done. A day worksheet is used to record work done that wasn't originally planned and shown in the jobsheet.

P. Gresford Building Contractors

Day worksheet

Customer _Chris MacFarlane_   **Date** _____

Description of work being carried out _____

Repoint front of property.

_____

Labour	Craft	Hours	Gross rate	TOTALS
Materials	**Quantity**	**Rate**	**% addition**	
Plant	**Hours**	**Rate**	**% addition**	

Comments

Signed _____   **Date** _____

Site manager/foreman signature _____

Figure 1.6 A typical day worksheet

Requisition form

A requisition form (also known as an order form) is used when you require plant, materials or equipment. Once you have worked out what you need and how much of it you need, a requisition form can then be filled in and sent to the relevant supplier.

Remember

Make sure you have all the tools and equipment you need before you go to do a job. You will need to plan ahead and fill in a requisition form early!

P. Gresford Building Contractors

Requisition form

Supplier _____ Order no. _____
_____ Serial no. _____
Tel no. _____ Contact _____
Fax no. _____ Our ref _____

Contract/Delivery address/Invoice address   Statements/applications
_____   for payments to be sent to
_____   _____
Tel no. _____   _____
Fax no. _____   _____

Item no.	Quantity	Unit	Description	Unit price	Amount

Total £ _____

Date _____

Payment terms _____
Originated by _____
Authorised by _____

Figure 1.7 A typical requisition form (order form)

Delivery note

A delivery note is sent by a supplier along with an order. It lists the materials delivered and the quantity. If you receive a delivery, you must check the delivery note against the tools, equipment or materials delivered. If everything matches, then you can sign the note. If anything is missing or damaged, you should not sign the note and must inform your supervisor.

Delivery note

Bailey & Sons Ltd

Building materials supplier

Tel: 01234 567890

Your ref: AB00671

Our ref: CT020   **Date:** 17 Jul 2006

Order no: 67440387

Invoice address: **Delivery address:**
Carillion Training Centre,   Same as invoice
Deptford Terrace, Sunderland

Description of goods	Quantity	Catalogue no.
OPC 25kg	10	OPC1.1

Comments:

Date and time of receiving goods:

Name of recipient (caps):

Signature:

Figure 1.8 A typical delivery note

Work programme

A work programme is a method of showing very easily what work is being carried out on a building and when. Used by many site agents or supervisors, a work programme is a bar chart that lists the tasks that need to be done down the left side and shows a timeline across the top (see Figure 1.9). A work programme is used to make sure that the relevant trade is on site at the correct time and that materials are delivered when needed. A site agent or supervisor can quickly tell from looking at the chart if work is keeping to schedule or falling behind.

		Time in days						
		1	2	3	4	5	6	7
Activity	A	■						
	B	■	■	■				
	C		■	■				
	D			■	■	■		
	E		■	■	■	■	■	■
	F				■	■	■	■
	G						■	■

Figure 1.9 A work programme

Getting involved in the construction industry

There are many ways of entering the construction industry, but the most common way is as an apprentice.

Apprenticeships

You can become an apprentice by:

1. Being employed directly by a construction company who will send you to college.

2. Being employed by a training provider, such as Carillion, which combines construction training with practical work experience.

An apprenticeship will give you on-the-job training and experience

On 1 August 2002, the construction industry introduced a mandatory induction programme for all apprentices joining the industry. The programme has four distinct areas:

Chapter 1 The construction industry

1. apprenticeship framework requirements
2. the construction industry
3. employment
4. health and safety.

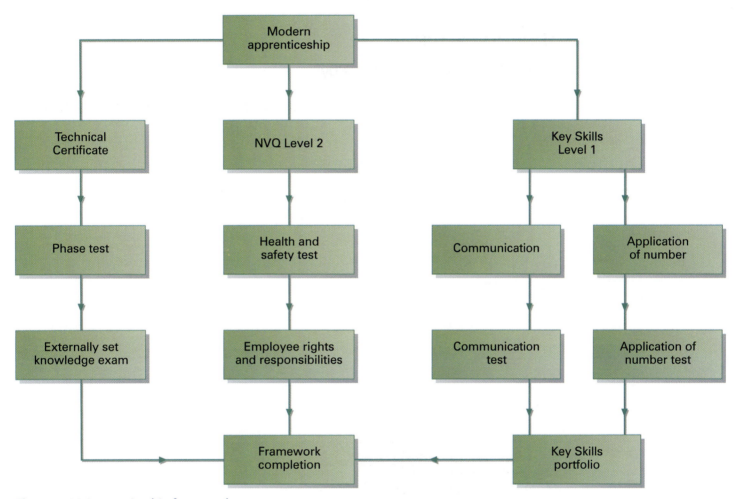

Figure 1.10 Apprenticeship framework

Apprenticeship frameworks are based on a number of components designed to prepare people for work in a particular construction occupation.

Construction frameworks are made up of the following mandatory components:

- NVQs
- technical certificates (construction awards)
- key skills.

However, certain trades require additional components. Bricklaying, for example, requires abrasive wheels certification.

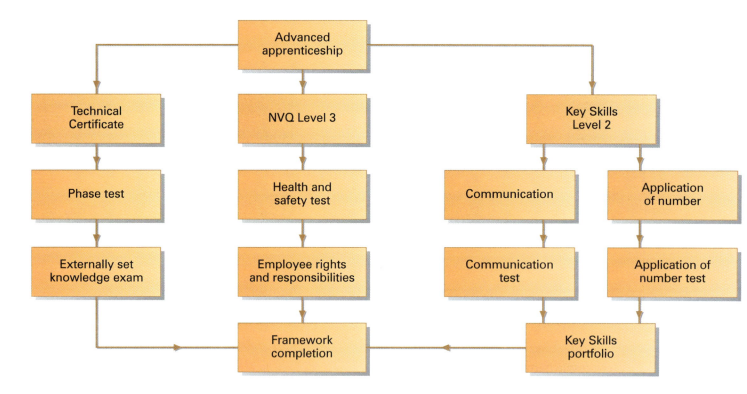

Figure 1.11 Advanced apprenticeship framework

National Vocational Qualifications (NVQs)

NVQs are available to anyone, with no restrictions on age, length or type of training, although learners below a certain age can only perform certain tasks. There are different levels of NVQ (e.g. 1, 2, 3), which in turn are broken down into units of competence. NVQs are not like traditional examinations in which someone sits an exam paper. An NVQ is a 'doing' qualification, which means it lets the industry know that you have the knowledge, skills and ability to actually 'do' something.

The Construction Industry Training Board (CITB) is the national training organisation for construction in the UK and is responsible for setting training standards. NVQs are made up of both mandatory and optional units and the number of units that you need to complete for an NVQ depends on the level and the occupation.

NVQs are assessed in the workplace, and several types of evidence are used:

- Witness testimony consists of evidence provided by various individuals who have firsthand knowledge of your work and performance relating to the NVQ. Work colleagues, supervisors and even customers can provide evidence of your performance.
- Your natural performance can be observed a number of times in the workplace while carrying out work-related activities.
- The use of historical evidence means that you can use evidence from past achievements or experience, if it is directly related to the NVQ.
- Assignments or projects can be used to assess your knowledge and understanding of a subject.
- Photographic evidence showing you performing various tasks in the workplace can be used, providing it is authenticated by your supervisor.

Technical certificates

Technical certificates are often related to NVQs. A certificate provides evidence that you have the underpinning knowledge and understanding required to complete a particular task. An off-the-job training programme, either in a college or with a training provider, may deliver technical certificates. You generally have to sit an end-of-programme exam to achieve the full certificate.

Key skills

Some students have key skills development needs, so learners and apprentices must achieve key skills at Level 1 or 2 in both Communications and Application of number. Key skills are signposted in each level of the NVQ and are assessed independently, so you will need to be released from your training to attend a key skills test.

Employment

Conditions of employment are controlled by legislation and regulations. The Department of Trade and Industry (DTI) publishes most of this legislation. To find out more about your working rights, visit the DTI website. A quick link has been made available at www.heinemann.co.uk/hotlinks – just enter the express code 0866P.

The main pieces of legislation that will apply to you are:

- The Employment Act 2002 which gives extra rights to working parents and gives new guidance on resolving disputes, amongst other things.
- The Employment Relations Act 1999 covers areas such as trade union membership and disciplinary and grievance proceedings.

The Race Relations Act protects people of all skin colours, races and nationalities

- The Employment Rights Act 1996 details the rights an employee has by law, including the right to have time off work and the right to be given notice if being dismissed.
- The Sex Discrimination Acts of 1975 and 1986 state that it is illegal for an employee to be treated less favourably because of their sex, for example, paying a man more than a woman or offering a woman more holiday than a man, even though they do the same job.
- The Race Relations Act 1976 states that it is against the law for someone to be treated less favourably because of their skin colour, race, nationality or ethnic origin.

- The Disability Discrimination Act 1995 makes it illegal for someone to be treated less favourably just because they have a physical or mental disability.
- The National Minimum Wage Act 1998 makes sure that everyone in the UK is paid a minimum amount. How much you must be paid depends on how old you are and whether or not you are on an Apprenticeship Scheme. The national minimum wage is periodically assessed and increased so it is a good idea to make sure you know what it is. At the time of writing, under 18s and those on Apprenticeship Schemes do not qualify for the minimum wage. For those aged 18–21, the minimum wage is £4.25 per hour and for adult workers aged 22 or over, the minimum wage is £5.05 per hour.

Find out

What is the national minimum wage at the moment? You can find out from lots of different places, including the DTI website. You can find a link to the site at www.heinemann.co.uk/hotlinks – just enter the express code 0866P.

Contract of employment

Within two months of starting a new job, your employer must give you a contract of employment. This will tell you the terms of your employment and should include the following information:

- job title
- place of work
- hours of work
- rates of pay
- holiday pay
- overtime rates
- statutory sick pay
- pension scheme
- discipline procedure
- termination of employment
- dispute procedure.

If you have any questions about information contained within your contract of employment, you should talk to your supervisor before you sign it.

When you start a new job, you should also receive a copy of the safety policy and an employee handbook containing details of the general policy, procedures and disciplinary rules.

Discrimination in the workplace

Discrimination means treating someone unjustly, and in the workplace it can range from bullying, intimidation or harassment to paying someone less money or not giving them a job. Discriminating against people within the working environment is against the law. This includes discrimination on the grounds of:

- sex, gender or sexual orientation
- race, colour, nationality or ethnic origin
- religious beliefs
- disability.

The law states that employment, training and promotion should be open to all employees regardless of any of the above. Pay should be equal for men and women if they are required to do the same job.

Men and women must be treated equally at work

Sources of information and advice

There are many places you can go to get information and advice about a career in the construction industry. If you are already studying, you can speak to your tutor, your school or college careers advisor or you can get in touch with Connexions for careers advice especially for young people. Visit www.heinemann.co.uk/hotlinks and enter the express code 0866P for a link to Connexions' website. You can also find their telephone number in your local phonebook.

Organisations such as those listed below are very good sources of careers advice specific to the construction industry.

- CITB (Construction Industry Training Board) – the industry's national training organisation
- City and Guilds – a provider of recognised vocational qualifications
- The Chartered Institute of Building Services Engineers
- The Institute of Civil Engineers
- Trade unions such as GMB (Britain's General Union), UCATT (Union of Construction, Allied Trades and Technicians), UNISON (the public services union), Amicus (the manufacturing union, previously MSF)

Links to all these organisations' websites can be found by visiting www.heinemann.co.uk/hotlinks and entering the express code 0866P.

FAQ

Why do I need to learn about different trades?

It is very important that you have some basic knowledge of what other trades do. This is because you will often work with people from other trades and their work will affect yours and vice versa.

What options do I have once I have gained my NVQ Level 2 qualification?

Once you are qualified, there is a wide range of career opportunities available to you. For example, you could progress from a tradesman to a foreman and then to a site agent. There may also be the opportunity to become a clerk of works, an architect or a college lecturer. Some tradesmen are happy to continue as tradesmen and some start up their own businesses.

Knowledge check

1. How many members of staff are there in a small company, a medium company and a large company?
2. Give an example of a public construction project. Who pays for public work?
3. Name a job in each of the four construction employment areas: professional; technician; building craft worker; building operative.
4. Why is the client the most important member of the building team?
5. Explain the meaning of the following building types: a) residential, b) low rise, and c) semi-detached.
6. What is the substructure of a building?
7. What are the three different methods of communication?
8. What information might a schedule give you?
9. What does NVQ stand for?
10. What information must be in your contract of employment?

chapter 2

Health and safety

OVERVIEW

Every year in the construction industry over 100 people are killed and thousands more are seriously injured as a result of the work that they do. There are thousands more who suffer from health problems, such as dermatitis, asbestosis, industrial asthma, vibration white finger and deafness. You can therefore see why learning as much as you can about health and safety is very important.

This chapter will cover:

- Health and safety legislation
- Health and welfare in the construction industry
- Manual handling
- Fire and fire-fighting equipment
- Safety signs
- Personal protective equipment (PPE)
- Reporting accidents
- Risk assessment.

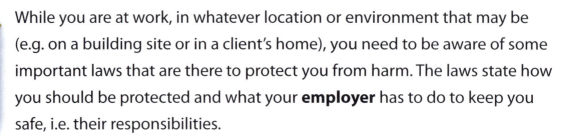

Health and safety legislation

While you are at work, in whatever location or environment that may be (e.g. on a building site or in a client's home), you need to be aware of some important laws that are there to protect you from harm. The laws state how you should be protected and what your **employer** has to do to keep you safe, i.e. their responsibilities.

Health and safety legislation not only protects you, but also states what your responsibilities are in order to keep others safe. It is very important that you follow any guidance given to you regarding health and safety and that you know what your responsibilities are.

What is legislation?

The word legislation generally refers to a law that is made in Parliament and is often called an act. For our purposes, health and safety acts state what should and shouldn't be done by employers and employees in order to keep work places safe. If an employer or an employee does something they shouldn't, or just as importantly, doesn't do something they should, they could face paying a large fine or even a prison sentence.

Health and safety legislation you need to be aware of

There are a lot of different pieces of legislation and regulations that affect the construction industry. Over the next few pages are just a few of those that you need to be aware of. Some of these are dealt with in more detail later on in this chapter.

Health and Safety at Work Act 1974

The Health and Safety at Work Act 1974 applies to all places of work, not just construction environments. It not only protects employers and employees but also any member of the public who might be affected by the work being done. The act outlines what must be done by employers and employees to ensure that the work they do is safe.

Definition

Employer – the person or company you work for

Did you know?

The average fine for breaking a health and safety law in the year 2003/04 was £9,858. The largest fine was £700,000

The main objectives of the Health and Safety at Work Act are:

- To ensure the health, safety and welfare of all persons at work.
- To protect the general public from work activities.
- To control the use, handling, storage and transportation of explosives and highly flammable substances.
- To control the release of **noxious** or offensive substances into the atmosphere.

The Health and Safety at Work Act is **enforced** by the **Health and Safety Executive** (HSE). HSE inspectors have the power to:

- Enter any premises to carry out investigations.
- Take statements and check records.
- Demand seizure, dismantle, neutralise or destroy anything that is likely to cause immediate serious injury.
- Issue an improvement notice, which gives a company a certain amount of time to sort out a health and safety problem.
- Issue a prohibition notice, which stops all work until the situation is safe.
- Give guidance and advice on health and safety matters.
- **Prosecute** people who break the law, including employers, employees, self-employed manufacturers and suppliers.

As we learnt at the beginning of this chapter, employers and employees have certain responsibilities under health and safety legislation. These are often referred to as 'duties' and are things that should or shouldn't be done by law. If you do not carry out your duties, you are breaking the law and you could be prosecuted.

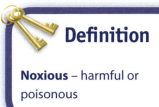

Definition

Noxious – harmful or poisonous

Find out

Will you be working with any highly flammable, explosive or noxious substances? What are they?

Definition

Enforced – making sure a law is obeyed

Prosecute – to accuse someone of committing a crime, which usually results in being taken to court and, if found guilty, being punished

Duties of the employer

Under the Health and Safety at Work Act employers must:

- provide a safe entrance and exit to the workplace
- provide a safe place to work
- provide and maintain safe machinery and equipment
- provide employees with the necessary training to be able to do their job safely
- have a written safety policy
- ensure safe handling, transportation and storage of machinery, equipment and materials
- provide personal protective equipment (PPE)
- involve trade union safety representatives, where appointed, in all matters relating to health and safety.

You have a legal duty to work safely at all times

Duties of the employee

Under the Health and Safety at Work Act employees must:

- take care at all times and ensure that they do not put themselves or others at risk by their actions
- co-operate with employers in regard to health and safety
- use any equipment and safeguards provided by their employer
- not misuse or interfere with anything that is provided for their safety.

Control of Substances Hazardous to Health Regulations 2002 (COSHH)

The COSHH regulations state how employees and employers should work with, handle, move and dispose safely of potentially dangerous substances. A substance hazardous to health is anything that might negatively affect your health, for example:

- dust or small particles from things like bricks and wood and fumes from chemicals
- chemicals in things like paint, **adhesives** and cement
- explosive or flammable chemicals or material.

The main aim of the COSHH regulations is to ensure that any risks due to working with hazardous substances or being exposed to them are assessed. Action must then be taken to eliminate or control the risks.

There are three different ways in which hazardous substances can enter the body:

1. Inhalation – breathing in the dangerous substance
2. Absorption – when the hazardous substance enters the body through the skin
3. Ingestion – taking in the hazardous substance through the mouth.

The COSHH regulations are as follows:

1. You should know exactly what products and substances you are using. You should be told this information by your employer.
2. Any hazards to health from using a substance or being exposed to it must be assessed by your employer.
3. If a substance is associated with any hazards to health, your employer must eliminate or control the hazard by either using a different substance or by making sure the substance is used according to guidelines (i.e. used outside or only used for short periods of time). Your

Definition

Adhesive – glue

Find out

Will you be working with any substances hazardous to health? What precautions and safety measures do you think should be taken for each?

Remember

It is not always possible to see a harmful substance so, if you are given any PPE or instructions about how to use/move/dispose of something, use them. Don't think that just because you can't see a hazardous substance, it isn't there

employer must also provide you with appropriate PPE and make sure that all possible precautions are taken.

4. Your employer must ensure that people are properly trained and informed of any hazards. All staff should be trained to recognise identifiable hazards and should know the correct precautions to take.

5. In order to make sure precautions are up to date, your employer has to monitor all tasks and change any control methods when required.

6. In case anyone ever needs to know what happened in the past, a record of all substances used by employees must be kept.

Provision and Use of Work Equipment Regulations 1998 (PUWER)

The PUWER regulations cover all working equipment such as tools and machinery. Under the PUWER regulations, employers must make sure that any tools and equipment they provide are:

- suitable for the job
- maintained (serviced and repaired)
- inspected (a regular check that ensures the piece of equipment and its parts are still in good working condition).

Safety tip

If you come across a substance that you are unsure about, do not use it. Report it to your supervisor as soon as possible

Under PUWER, all tools and equipment must be regularly serviced and repaired

Employers also have to make sure that any risk of harm from using the equipment has been identified and all precautions and safety measures have been taken. Employers must also ensure that anyone who uses the equipment has been properly trained and instructed in how to do so.

The Manual Handling Operations Regulations 1992

These regulations cover all work activities in which a person does the lifting instead of a machine. The correct and safe way to lift, which reduces the risk of injury, is covered later on in this chapter (see page 00).

The Control of Noise at Work Regulations 2005

In the course of your career in construction, it is likely that you will be at some time working in a noisy environment. The Control of Noise at Work Regulations are there to protect you against the consequences of being exposed to high levels of noise, which can lead to permanent hearing damage.

Damage to hearing can be caused by:

- the volume of noise (measured in decibels)
- the length of time exposed to the noise (over a day, over a lifetime etc.).

The regulations give guidance on the maximum period of time someone can be safely exposed to a decibel level, and your employer has to follow it.

If you have access to the internet, you might wish to visit the Health and Safety Executive website and find out what it is like to have hearing loss caused by long-term exposure to noise. A link to the web page has been made available at www.heinemann.co.uk/hotlinks – just enter the express code 0866P.

The Work at Height Regulations 2005

It is not at all unusual for a construction worker to carry out their everyday job high up off the ground, for example, on scaffolding, on a ladder, or on the roof of a building. The Work at Height Regulations make sure that your employer does all that they can to reduce the risk of injury or death from working at height. Your employer has a duty to:

- avoid work at height where possible
- use equipment that will prevent falls
- use equipment and other methods that will minimise the distance and consequences of a fall.

As an employee, under the regulations you must follow any training that has been given to you, report any hazards to your supervisor and use any safety equipment that is made available to you.

The Electricity at Work Regulations 1989

The Electricity at Work Regulations cover any work that involves the use of electricity or electrical equipment. Your employer has a duty to make sure that electrical systems you may come into contact with are safe and regularly maintained. They also have to make sure that they have done everything the law states to reduce the risk of an employee coming into contact with a live electrical current.

The Personal Protective Equipment at Work Regulations 1992

There are certain situations in which you will need to wear personal protective equipment (PPE). The Personal Protective Equipment at Work Regulations details the different types of PPE that are available and states when they should be worn. Your employer has to ensure appropriate PPE is available for certain tasks (e.g. gloves when working with solvents, face masks when cutting bricks, safety goggles when using a circular saw).

The different types of PPE available are covered in more detail later on in this chapter (see page 57).

Reporting of Injuries, Diseases and Dangerous Occurrences Regulations 1995 (RIDDOR)

Employers have duties under RIDDOR to report accidents, diseases or dangerous occurrences. This information is used by the HSE to identify where and how risk arises and to investigate serious accidents.

Several other regulations exist which cover very specific things such as asbestos, pressure equipment and lead paint. If you want to find out more about these regulations, or any others, ask your tutor or employer for more information or visit the Health and Safety Executive website (go to www.heinemann.co.uk/hotlinks and enter the express code 0866P for a quick link).

Health and welfare in the construction industry

Jobs in the construction industry have one of the highest injury and accident rates and as a worker you will be at constant risk unless you adopt a good health and safety attitude. By following the rules and regulations set out to protect you and by taking reasonable care of yourself and others, you will become a safe worker and thus reduce the chance of any injuries or accidents.

Remember

Health and safety laws are there to protect you and other people. If you take shortcuts or ignore the rules, you are placing yourself and others at serious risk.

The most common risks to a construction worker

What do you think these might be? Think about the construction industry you are working in and the hazards and risks that exist.

The most common health and safety risks a construction worker faces are:

- accidents
- ill health.

Accidents

We often hear the saying 'accidents will happen', but when working in the construction industry, we should not accept that accidents just happen

Accidents can happen if your work area is untidy

sometimes. When we think of an accident, we quite often think about it as being no-one's fault and something that could not have been avoided. The truth is that most accidents are caused by human error, which means someone has done something they shouldn't have done or, just as importantly, not done something they should have done.

Accidents often happen when someone is hurrying, not paying enough attention to what they are doing or they have not received the correct training.

If an accident happens, you or the person it happened to may be lucky and will not be injured. More often, an accident will result in an injury which may be minor (e.g. a cut or a bruise) or possibly major (e.g. loss of a limb). Accidents can also be fatal. The most common causes of fatal accidents in the construction industry are:

- falling from scaffolding
- being hit by falling objects and materials
- falling through fragile roofs
- being hit by forklifts or lorries
- electrocution.

Chapter 2 Health and safety

Ill health

While working in the construction industry, you will be exposed to substances or situations that may be harmful to your health. Some of these health risks may not be noticeable straight away and it may take years for **symptoms** to be noticed and recognised.

Ill health can result from:

- exposure to dust (such as asbestos), which can cause breathing problems and cancer
- exposure to solvents or chemicals, which can cause **dermatitis** and other skin problems
- lifting heavy or difficult loads, which can cause back injury and pulled muscles
- exposure to loud noise, which can cause hearing problems and deafness
- using vibrating tools, which can cause **vibration white finger** and other problems with the hands.

Everyone has a responsibility for health and safety in the construction industry but accidents and health problems still happen too often. Make sure you do what you can to prevent them.

Definition

Symptom – a sign of illness or disease (e.g. difficulty breathing, a sore hand or a lump under the skin)

Definition

Dermatitis – a skin condition where the affected area is red, itchy and sore

Vibration white finger – a condition that can be caused by using vibrating machinery (usually for very long periods of time). The blood supply to the fingers is reduced which causes pain, tingling and sometimes spasms (shaking)

Staying healthy

As well as keeping an eye out for hazards, you must also make sure that you look after yourself and stay healthy. One of the easiest ways to do this is to wash your hands on a regular basis. By washing your hands you are preventing hazardous substances from entering your body through ingestion (swallowing). You should always wash your hands after going to the toilet and before eating or drinking.

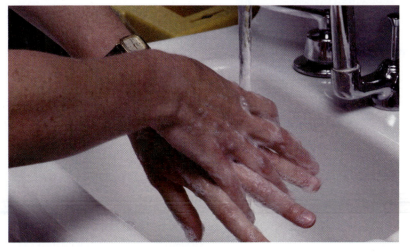

Always wash your hands to prevent ingesting hazardous substances

43

Definition

Barrier cream – a cream used to protect the skin from damage or infection

Definition

Corrosive – a substance that can damage things it comes into contact with (e.g. material, skin)

Toxic – poisonous

Contamination – when harmful chemicals or substances pollute something (e.g. water)

Other precautions that you can take are ensuring that you wear **barrier cream**, the correct PPE and only drink water that is labelled as drinking water. Remember that some health problems do not show symptoms straight away and what you do now can affect you much later in life.

Welfare facilities

Welfare facilities are things such as toilets, which must be provided by your employer to ensure a safe and healthy workplace. There are several things that your employer must provide to meet welfare standards and these are:

- Toilets – the number of toilets provided depends upon the number of people who are intended to use them. Males and females can use the same toilets providing there is a lock on the inside of the door. Toilets should be flushable with water or, if this is not possible, with chemicals.

- Washing facilities – employers must provide a basin large enough to allow people to wash their hands, face and forearms. Washing facilities must have hot and cold running water as well as soap and a means of drying your hands. Showers may be needed if the work is very dirty or if workers are exposed to **corrosive** and **toxic** substances.

- Drinking water – there should be a supply of clean drinking water available, either from a tap connected to the mains or from bottled water. Taps connected to the mains need to be clearly labelled as drinking water and bottled drinking water must be stored in a separate area to prevent **contamination**.

- Storage or dry room – every building site must have an area where workers can store the clothes that they do not wear on site, such as coats and motorcycle helmets. If this area is to be used as a drying room then adequate heating must also be provided in order to allow clothes to dry.

- Lunch area – every site must have facilities that can be used for taking breaks and lunch well away from the work area. These facilities must provide shelter from the wind and rain and be heated as required. There

Chapter 2 Health and safety

should be access to tables and chairs, a kettle or urn for boiling water and a means of heating food, such as a microwave.

When working in an occupied house, you should make arrangements with the client to use the facilities in their house.

Safety tip

When placing clothes in a drying room, do not place them directly onto heaters as this can lead to fire

On the job: Scaffolding safety

Ralph and Vijay are working on the second level of some scaffolding clearing debris. Ralph suggests that, to speed up the task, they should throw the debris over the edge of the scaffolding into a skip below. The building Ralph and Vijay are working on is on a main road and the skip is not in a closed off area. What do you think of Ralph's idea? What are your reasons for this answer?

Manual handling

Manual handling means lifting and moving a piece of equipment or material from one place to another without using machinery. Lifting and moving loads by hand is one of the most common causes of injury at work. Most injuries caused by manual handling result from years of lifting items that are too heavy, are awkward shapes or sizes, or from using the wrong technique. However, it is also possible to cause a lifetime of back pain with just one single lift.

Poor manual handling can cause injuries such as muscle strain, pulled ligaments and hernias. The most common injury by far is spinal injury. Spinal injuries are very serious because there is very little that doctors can do to correct them and, in extreme cases, workers have been left paralysed.

45

Poor manual handling techniques can lead to serious permanent injury

What you can do to avoid injury

The first and most important thing you can do to avoid injury from lifting is to receive proper manual handling training. Kinetic lifting is a way of lifting objects that reduces the chance of injury and is covered in more detail on the next page.

Before you lift anything you should ask yourself some simple questions:

- Does the object need to be moved?

- Can I use something to help me lift the object? A mechanical aid such as a forklift or crane or a manual aid such as a wheelbarrow may be more appropriate than a person.

- Can I reduce the weight by breaking down the load? Breaking down a load into smaller and more manageable weights may mean that more journeys are needed, but it will also reduce the risk of injury.

- Do I need help? Asking for help to lift a load is not a sign of weakness and team lifting will greatly reduce the risk of injury.

- How much can I lift safely? The recommended maximum weight a person can lift is 25 kg but this is only an average weight and each person is different. The amount that a person can lift will depend on their physique, age and experience.

- Where is the object going? Make sure that any obstacles in your path are out of the way before you lift. You also need to make sure there is somewhere to put the object when you get there.

- Am I trained to lift? The quickest way to receive a manual handling injury is to use the wrong lifting technique.

Lifting correctly (kinetic lifting)

When lifting any load it is important to keep the correct posture and to use the correct technique.

The correct posture before lifting:

- feet shoulder width apart with one foot slightly in front of the other
- knees should be bent
- back must be straight
- arms should be as close to the body as possible
- grip must be firm using the whole hand and not just the finger tips.

The correct technique when lifting:

- approach the load squarely facing the direction of travel
- adopt the correct posture (as above)
- place hands under the load and pull the load close to your body
- lift the load using your legs and not your back.

When lowering a load you must also adopt the correct posture and technique:

- bend at the knees, not the back
- adjust the load to avoid trapping fingers
- release the load.

Remember

Even light loads can cause back problems so, when lifting anything, always take care to avoid twisting or stretching

Brickwork NVQ and Technical Certificate Level 2

Think before lifting

Adopt the correct posture before lifting

Get a good grip on the load

Chapter 2 Health and safety

Adopt the correct posture when lifting

Move smoothly with the load

Adopt the correct posture and technique when lowering

49

Fire and fire-fighting equipment

Fires can start almost anywhere and at any time but a fire needs three things to burn. These are:

1. fuel
2. heat
3. oxygen.

This can be shown in what is known as 'the triangle of fire'. If any of the sides of the triangle are removed, the fire cannot burn and it will go out.

Find out

What fire risks are there in the construction industry? Think about some of the materials (fuel) and heat sources that could make up two of the sides of 'the triangle of fire'

Figure 2.1 The triangle of fire

Remember:

- Remove the fuel and there is nothing to burn so the fire will go out.
- Remove the heat and the fire will go out.
- Remove the oxygen and the fire will go out as fire needs oxygen to survive.

Fires can be classified according to the type of material that is involved:

- Class A – wood, paper, textiles etc.
- Class B – flammable liquids, petrol, oil etc.
- Class C – flammable gases, liquefied petroleum gas (**LPG**), propane etc.
- Class D – metal, metal powder etc.
- Class E – electrical equipment.

Fire-fighting equipment

There are several types of fire-fighting equipment, such as fire blankets and fire extinguishers. Each type is designed to be the most effective at putting out a particular class of fire and some types should never be used in certain types of fire.

Fire extinguishers

A fire extinguisher is a metal canister containing a substance that can put out a fire. There are several different types and it is important that you learn which type should be used on specific classes of fires. This is because if you use the wrong type, you may make the fire worse or risk severely injuring yourself.

Fire extinguishers are now all one colour (red) but they have a band of colour which shows what substance is inside.

Water

The coloured band is red and this type of extinguisher can be used on Class A fires. Water extinguishers can also be used on Class C fires in order to cool the area down.

A water fire extinguisher should NEVER be used to put out an electrical or burning fat/oil fire. This is because electrical current can carry along the jet of water back to the person holding the extinguisher, electrocuting them. Putting water

Water fire extinguisher

Brickwork NVQ and Technical Certificate Level 2

Foam fire extinguisher

Dry powder extinguisher

on to burning fat or oil will make the fire worse as the fire will 'explode', potentially causing serious injury.

Foam

The coloured band is cream and this type of extinguisher can also be used on Class A fires. A foam extinguisher can also be used on a Class B fire if the liquid is not flowing and on a Class C fire if the gas is in liquid form.

Carbon dioxide (CO_2)

The coloured band is black and the extinguisher can be used on Class A, B, C and E fires.

Dry powder

The coloured band is blue and this type of extinguisher can be used on all classes of fire. The powder puts out the fire by knocking down the flames.

Carbon dioxide (CO_2) extinguisher

Fire blankets

Fire blankets are normally found in kitchens or canteens as they are good at putting out cooking fires. They are made of a fireproof material and work by smothering the fire and stopping any more oxygen from getting to it, thus putting it out. A fire blanket can also be used if a person is on fire.

It is important to remember that when you put out a fire with a fire blanket, you need to take extra care as you will have to get quite close to the fire.

Chapter 2 Health and safety

A fire blanket

What to do in the event of a fire

During **induction** to any workplace, you will be made aware of the fire procedure as well as where the fire assembly points (also known as **muster points**) are and what the alarm sounds like. On hearing the alarm you must stop what you are doing and make your way to the nearest muster point. This is so that everyone can be accounted for. If you do not go the muster point or if you leave before someone has taken your name, someone may risk their life to go back into the fire to get you.

When you hear the alarm, you should not stop to gather any belongings and you must not run. If you discover a fire, you must only try to fight the fire if it is blocking your exit or if it is small. Only when you have been given the all-clear can you re-enter the site or building.

Remember

Fire and smoke can kill in seconds so think and act clearly, quickly and sensibly

Definition

Induction – a formal introduction you will receive when you start any new job, where you will be shown around, shown where the toilets and canteen etc. are, and told what to do if there is a fire

53

Safety signs

Safety signs can be found in many areas of the workplace and they are put up in order to:

- warn of any **hazards**
- prevent accidents
- inform where things are
- tell you what to do in certain areas.

Types of safety sign

There are many different safety signs but each will usually fit into one of four categories:

Definition

Hazard – a danger or risk

Figure 2.2 A prohibition sign

1. Prohibition signs – these tell you that something MUST NOT be done. They always have a white background and a red circle with a red line through it.

Chapter 2 Health and safety

Figure 2.3 A mandatory sign

2. Mandatory signs – these tell you that something MUST be done. They are also circular but have a white symbol on a blue background.

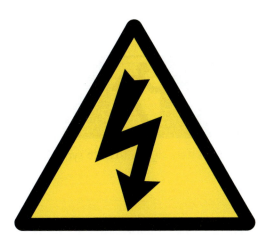

Figure 2.4 A warning sign

3. Warning signs – these signs are there to alert you to a specific hazard. They are triangular and have a yellow background and a black border.

55

Figure 2.5 An information sign

4. Information signs – these give you useful information like the location of things (e.g. a first aid point). They can be square or rectangular and are green with a white symbol.

Figure 2.6 A safety sign with both symbol and words

Most signs only have symbols that let you know what they are saying. Others have some words as well, for example, a no smoking sign might have a cigarette in a red circle, with a red line crossing through the cigarette and the words 'No smoking' underneath.

Chapter 2 Health and safety

Personal protective equipment (PPE)

Personal protective equipment (PPE) is a form of defence against accidents or injury and comes in the form of articles of clothing. This is not to say that PPE is the only way of preventing accidents or injury. It should be used together with all the other methods of staying healthy and safe in the workplace (i.e. equipment, training, regulations and laws etc.).

PPE must be supplied by your employer free of charge and you have responsibility as an employee to look after it and use it whenever it is required.

Types of PPE

There are certain parts of the body that require protection from hazards during work and each piece of PPE must be suitable for the job and used properly.

Head protection

There are several different types of head protection but the one most commonly used in construction is the safety helmet (or hard hat). This is used to protect the head from falling objects and knocks and has an adjustable strap to ensure a snug fit. Some safety helmets come with attachments for ear defenders or eye protection. Safety helmets are meant to be worn directly on the head and must not be worn over any other type of hat.

Remember

Make sure you take notice of safety signs in the workplace – they have been put up for a reason!

Remember

PPE only works properly if it is being used and used correctly!

A safety helmet

57

Eye protection

Eye protection is used to protect the eyes from dust and flying debris. The three main types are:

1. Safety goggles – made of a durable plastic and used when there is a danger of dust getting into the eyes or a chance of impact injury.

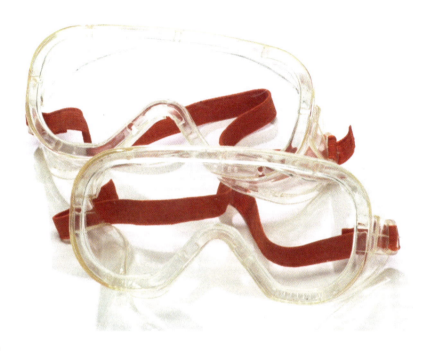

Safety goggles

2. Safety spectacles – these are also made from a durable plastic but give less protection than goggles. This is because they don't fully enclose the eyes and so only protect from flying debris.

Chapter 2 Health and safety

Safety spectacles

3. Facemasks – again made of durable plastic, facemasks protect the entire face from flying debris. They do not, however, protect the eyes from dust.

Foot protection

Safety boots or shoes are used to protect the feet from falling objects and to prevent sharp objects such as nails from injuring the foot. Safety boots should have a steel toe-cap and steel mid-sole.

Safety boots

Hearing protection

Hearing protection is used to prevent damage to the ears caused by very loud noise. There are several types of hearing protection available but the two most common types are earplugs and ear defenders.

Ear-plugs

1. Ear-plugs – these are small fibre plugs that are inserted into the ear and used when the noise is not too severe. When using ear-plugs, make sure that you have clean hands before inserting them and never use plugs that have been used by somebody else.

Ear defenders

2. Ear defenders – these are worn to cover the entire ear and are connected to a band that fits over the top of the head. They are used when there is excessive noise and must be cleaned regularly.

Chapter 2 Health and safety

Respiratory protection

Respiratory protection is used to prevent the worker from breathing in any dust or fumes that may be hazardous. The main type of respiratory protection is the dust mask.

Dust masks are used when working in a dusty environment and are lightweight, comfortable and easy to fit. They should be worn by only one person and must be disposed of at the end of the working day. Dust masks will only offer protection from non-toxic dust so, if the worker is to be exposed to toxic dust or fumes, a full respiratory system should be used.

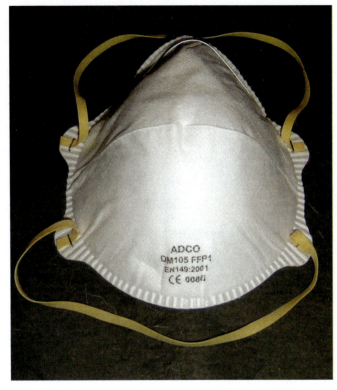

A dust mask

Hand protection

There are several types of hand protection and each type must be used for the correct task. For example, wearing lightweight rubber gloves to move glass will not offer much protection so leather gauntlets must be used. Plastic-coated gloves will protect you from certain chemicals and Kevlar® gloves offer cut resistance. To make sure you are wearing the most suitable type of glove for the task, you need to look first at what is going to be done and then match the type of glove to that task.

Safety gloves

61

Reporting accidents

When an accident occurs, there are certain things that must be done. All accidents need to be reported and recorded in the accident book and the injured person must report to a trained first aider in order to receive treatment. Serious accidents must be reported under the Reporting of Injuries, Diseases and Dangerous Occurrences Regulations 1995 (RIDDOR). Under RIDDOR your employer must report to the HSE any accident that results in:

- death
- major injury
- an injury that means the injured person is not at work for more than three consecutive days.

The accident book

The accident book is completed by the person who had the accident or, if this is not possible, someone who is representing the injured person.

The accident book will ask for some basic details about the accident, including:

- who was involved
- what happened
- where it happened
- the day and time of the accident
- any witnesses to the accident
- the address of the injured person
- what PPE was being worn
- what first aid treatment was given.

Chapter 2 Health and safety

Report of an Accident, Dangerous Occurrence or Near Miss

Date of incident _____ Time of incident _____

Location of incident _____

Details of person involved in accident

Name _____ Date of birth _____ Sex _____

Address _____

_____ Occupation _____

Date off work (if applicable) _____ Date returning to work _____

Nature of injury _____

Management of injury   ☐ First Aid only   ☐ Advised to see doctor
                        ☐ Sent to casualty ☐ Admitted to hospital

Account of accident, dangerous occurrence or near miss
(Continued on separate sheet if necessary)

Witnesses to the incident
(Names, addresses and occupations)

Was the injured person wearing PPE? If yes, what PPE? _____

Signature of person completing form _____

Occupation _____ Date _____

Figure 2.7 A typical accident book page

Definition

Proactive – taking action *before* something happens (e.g. an accident)

Reactive – taking action *after* something happens

As well as reporting accidents, 'near misses' must also be reported. This is because near misses are often the accidents of the future. Reporting near misses might identify a problem and can prevent accidents from happening in the future. This allows a company to be **proactive** rather than **reactive**.

63

Risk assessments

A risk assessment is where the dangers of an activity are measured against the likelihood of accidents taking place. People carry out risk assessments hundreds of times each day without even knowing it. For example, every time we cross the road we do a risk assessment without even thinking about it.

In the construction industry, risk assessments are done by experienced people who are able to identify what risks each task has. They are then able to put measures in place to control the risks they have identified. At some point in your career, you will have to carry out a risk assessment. You will be given proper training in how to do this but, until then, it is still important that you understand how risk assessments work. Below is an example of an everyday situation (crossing the road) and how a risk assessment would be carried out for this.

Step 1

Identify the hazards (the dangers) – in this situation the hazards are vehicles travelling at speed.

Step 2

Identify who will be at risk – the person crossing the road will be at risk, as will any drivers on the road who might have to swerve to avoid that person.

Step 3

Calculate the risk from the hazard against the likelihood of an accident taking place – the risk from the hazard is quite high because if an accident were to happen, the injury could be very serious. However, the likelihood of an accident happening is low because the chances of the person being hit while crossing are minimal.

Step 4

Introduce measures to reduce risk – in this case crossing the road at traffic lights or pedestrian crossings reduces risk.

Step 5

Monitor the risk – changes might need to be made to the risk assessment if there are any changes to the risks involved. In our example, changes might be traffic lights being out of order or an increase in the speed limit on the road.

On the job: Unidentified material

Craig and Kevin are clearing out a basement of all fixtures and fittings in preparation for a job. They uncover some white plaster-like material covering a heating pipe. They have never seen this type of material before and are not sure what it could be. Craig suggests that they just remove it so that they can get the job done. Kevin is not sure if this is such a good idea and wants to tell their boss about the material before they remove it. Who do you think has given the best suggestion? Why?

Knowledge check

1. Name five pieces of health and safety legislation that affect the construction industry.

2. What does HSE stand for? What does it do?

3. What does COSHH stand for?

4. What does RIDDOR stand for?

5. What might happen to you or your employer if a health and safety law is broken?

6. What are the two most common risks to construction workers?

7. State two things that you can do to avoid injury when lifting loads using manual handling techniques.

8. What three elements cause a fire and keep it burning?

9. What class(es) of fire can be put out with a carbon dioxide (CO_2) extinguisher?

10. What does a prohibition sign mean?

11. Describe how you would identify a warning sign.

12. Name the six different types of PPE.

13. Who fills in an accident report form?

14. Why is it important to report 'near misses'?

15. Briefly explain what a risk assessment is.

chapter 3
Working at height

OVERVIEW

Most construction trades require frequent use of some type of working platform or access equipment. Working off the ground can be dangerous and the greater the height the more serious the risk of injury. This chapter will give you a summary of some of the most common types of access equipment and provide information on how they should be used, maintained and checked to ensure that the risks to you and others are minimal.

This chapter will cover the following:

- General safety considerations
- Stepladders and ladders
- Roof work
- Trestle platforms
- Hop-ups
- Scaffolding

General safety considerations

You will need to be able to identify potential hazards associated with working at height, as well as hazards associated with equipment. It is essential that access equipment is well maintained and checked regularly for any deterioration or faults, which could compromise the safety of someone using the equipment and anyone else in the work area. Although obviously not as important as people, equipment can also be damaged by the use of faulty access equipment. When maintenance checks are carried out they should be properly recorded. This provides very important information that helps to prevent accidents.

Risk assessment

Before any work is carried out at height, a thorough risk assessment needs to be completed. Your supervisor or someone else more experienced will do this while you are still training, but it is important that you understand what is involved so that you are able to carry out an assessment in the future.

For a working at height risk assessment to be valid and effective a number of questions must be answered:

1. How is access and **egress** to the work area to be achieved?

2. What type of work is to be carried out?

3. How long is the work likely to last?

4. How many people will be carrying out the task?

5. How often will this work be carried out?

6. What is the condition of the existing structure (if any) and the surroundings?

7. Is adverse weather likely to affect the work and workers?

8. How competent are the workforce and their supervisors?

9. Is there a risk to the public and work colleagues?

> **Definition**
>
> **Egress** - an exit or way out

Duties

Your employer has a duty to provide and maintain safe plant and equipment, which includes scaffold access equipment and systems of work.

You have a duty:

- to comply with safety rules and procedures relating to access equipment

- to take positive steps to understand the hazards in the workplace and report things you consider likely to lead to danger, for example a missing handrail on a working platform

- not to tamper with or modify equipment.

Stepladders and ladders

Stepladders

A stepladder has a prop, which when folded out allows the ladder to be used without having to lean it against something. Stepladders are one of the most frequently used pieces of access equipment in the construction industry and are often used every day. This means that they are not always treated with the respect they demand. Stepladders are often misused – they should only be used for work that will take a few minutes to complete. When work is likely to take longer than this, a sturdier alternative should be found.

When stepladders are used, the following safety points should be observed:

- Ensure the ground on which the stepladder is to be placed is firm and level. If the ladder rocks or sinks into the ground it should not be used for the work.

- Always open the steps fully.

- Never work off the top tread of the stepladder.

- Always keep your knees below the top tread.

- Never use stepladders to gain additional height on another working platform.

> **Did you know?**
>
> Only a fully trained and competent person is allowed to erect any kind of working platform or access equipment. You should therefore not attempt to erect this type of equipment unless this describes you!

- Always look for the kitemark, which shows that the ladder has been made to British Standards.

Figure 3.1 British Standards Institution Kitemark

A number of other safety points need to be observed depending on the type of stepladder being used.

Wooden stepladder

Before using a wooden stepladder:

- Check for loose screws, nuts, bolts and hinges.
- Check that the tie ropes between the two sets of **stiles** are in good condition and not frayed.
- Check for splits or cracks in the stiles.
- Check that the treads are not loose or split.

Never paint any part of a wooden stepladder as this can hide defects, which may cause the ladder to fail during use, causing injury.

Definition

Stiles – the side pieces of a stepladder into which the steps are set

Wooden stepladder

Chapter 3 Working at height

Aluminium stepladder

Before using an aluminium stepladder:

- Check for damage to stiles and treads to see whether they are twisted, badly dented or loose.
- Avoid working close to live electricity supplies as aluminium will conduct electricity.

Aluminium stepladder

 Safety tip

If any faults are revealed when checking a wooden stepladder, it should be taken out of use, reported to the person in charge and a warning notice attached to it to stop anyone using it

Find out

What are the advantages and disadvantages of each type of stepladder?

 **Did you know?**

Stepladders should be stored under cover to protect from damage such as rust or rotting

Fibreglass stepladder

Before using a fibreglass stepladder, check for damage to stiles and treads. Once damaged, fibreglass stepladders cannot be repaired and must be disposed of.

Ladders

A ladder, unlike a stepladder, does not have a prop and so has to be leant against something in order for it to be used. Together with stepladders, ladders are one of the most common pieces of equipment used to carry out work at heights and gain access to the work area.

As with stepladders, ladders are also available in timber, aluminium and fibreglass and require similar checks before use.

 Safety tip

Ladders must NEVER be repaired once damaged and must be disposed of

Ladder types

Pole ladder

These are single ladders and are available in a range of lengths. They are most commonly used for access to scaffolding platforms. Pole ladders are made from timber and must be stored under cover and flat, supported evenly along their length to prevent them sagging and twisting. They should be checked for damage or defects every time before being used.

Pole ladder

Extension ladder

Extension ladders have two or more interlocking lengths, which can be slid together for convenient storage or slid apart to the desired length when in use.

Extension ladders are available in timber, aluminium and fibreglass. Aluminium types are the most favoured as they are lightweight yet strong and available in double and triple extension types. Although also very strong, fibreglass versions are heavy, making them difficult to manoeuvre.

Aluminium extension ladder

Erecting and using a ladder

The following points should be noted when considering the use of a ladder:

- As with stepladders, ladders are not designed for work of long duration. Alternative working platforms should be considered if the work will take longer than a few minutes.
- The work should not require the use of both hands. One hand should be free to hold the ladder.
- You should be able to do the work without stretching.
- You should make sure that the ladder can be adequately secured to prevent it slipping on the surface it is leaning against.

Pre-use checks

Before using a ladder check its general condition. Make sure that:

- no rungs are damaged or missing
- the stiles are not damaged
- no **tie-rods** are missing
- no repairs have been made to the ladder.

In addition, for wooden ladders ensure that:

- they have not been painted, which may hide defects or damage
- there is no decay or rot
- the ladder is not twisted or warped.

Erecting a ladder

Observe the following guidelines when erecting a ladder:

- Ensure you have a solid, level base.
- Do not pack anything under either (or both) of the stiles to level it.

Did you know?

On average in the UK, 14 people a year die at work falling from ladders; nearly 1200 suffer major injuries (source: Health and Safety Executive)

Definition

Tie-rods – metal rods underneath the rungs of a ladder that give extra support to the rungs

Remember

You must carry out a thorough risk assessment before working from a ladder. Ask yourself, 'Would I be safer using an alternative method?'

- If the ladder is too heavy to put it in position on your own, get someone to help.
- Ensure that there is at least a four-rung overlap on each extension section.
- Never rest the ladder on plastic guttering as it may break, causing the ladder to slip and the user to fall.
- Where the base of the ladder is in an exposed position, ensure it is adequately guarded so that no one knocks it or walks in to it.
- The ladder should be secured at both the top and bottom. The bottom of the ladder can be secured by a second person, however this person must not leave the base of the ladder whilst it is in use.
- The angle of the ladder should be a ratio of 1:4 (or 75°). This means that the bottom of ladder is 1 m away from the wall for every 4 m in height (see Figure 3.2).
- The top of the ladder must extend at least 1 m, or 5 rungs, above its landing point.

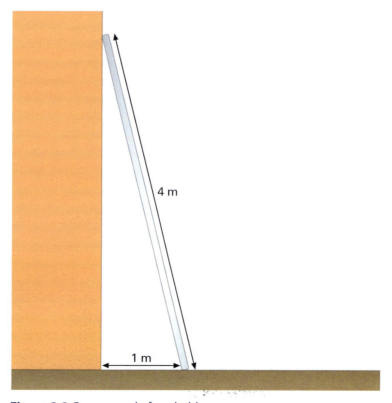

Figure 3.2 Correct angle for a ladder

Roof work

When carrying out any work on a roof, a roof ladder or **crawling board** must be used. Roof work also requires the use of edge protection or, where this is not possible, a safety harness.

Definition

Crawling board – a board or platform placed on roof joists which spread the weight of the worker allowing the work to be carried out safely

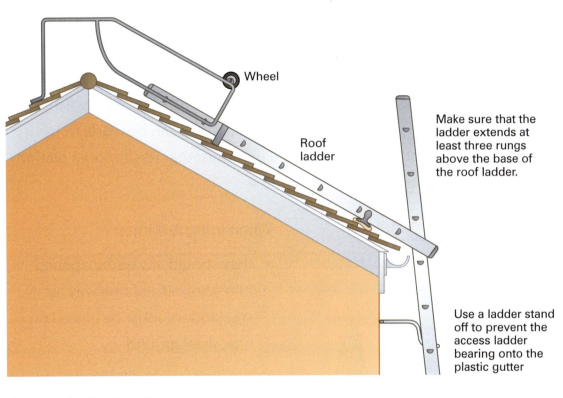

Figure 3.3 Roof work equipment

The roof ladder is rolled up the surface of the roof and over the ridge tiles, just enough to allow the ladder to be turned over and the ladder hook allowed to bear on the tiles on the other side of the roof. This hook prevents the roof ladder sliding down the roof once it is accessed.

Trestle platforms

A trestle is a frame upon which a platform or other type of surface (e.g. a table top) can be placed. A trestle should be used rather than a ladder for work that will take longer than a few minutes to complete. Trestle platforms are composed of the frame and the platform (sometimes called a stage).

Frames

A-frames

These are most commonly used by carpenters and painters. As the name suggests, the frame is in the shape of a capital A and can be made from timber, aluminium or fibreglass. Two are used together to support a platform (a scaffold or staging board). See Figure 3.4.

Safety tip

A-frame trestles should never be used as stepladders as they are not designed for this purpose

When using A-frames:

- they should always be opened fully and, in the same way as stepladders, must be placed on firm, level ground
- the platform width should be no less than 450 mm thick
- the overhang of the board at each end of the platform should be not more than four times its thickness.

Figure 3.4 A-frame trestles with scaffold board

Steel trestles

These are sturdier than A-frame trestles and are adjustable in height. They are also capable of providing a wider platform than timber trestles – see Figure 3.5. As with the A-frame type, they must be used only on firm and level ground but the trestle itself should be placed on a flat scaffold board on top of the ground. Trestles should not be placed more than 1.2 m apart.

Figure 3.5 Steel trestle with staging board

Platforms

Scaffold boards

To ensure that scaffold boards provide a safe working platform, before using them check that they:

- are not split
- are not twisted or warped
- have no large knots, which cause weakness.

Staging boards

These are designed to span a greater distance than scaffold boards and can offer a 600 mm wide working platform. They are ideal for use with trestles.

Safety tip

Do not use items as hop-ups that are not designed for the purpose (e.g. milk crates, stools or chairs). They are usually not very sturdy and can't take the weight of someone standing on them, which may result in falls and injury

Did you know?

It took 14 years of experimentation to finally settle on 48 mm as the diameter of most tubular scaffolding poles

Definition

Carded scaffolder – someone who holds a recognised certificate showing competence in scaffold erection

Hop-ups

Also known as step-ups, these are ideal for reaching low-level work that can be carried out in a relatively short period of time. A hop-up needs to be of sturdy construction and have a base of not less than 600 mm by 500 mm. Hop-ups have the disadvantage that they are heavy and awkward to move around.

Scaffolding

Tubular scaffold is the most commonly used type of scaffolding within the construction industry. There are two types of tubular scaffold:

1. Independent scaffold – free-standing scaffold that does not rely on any part of the building to support it (although it must be tied to the building to provide additional stability).

2. Putlog scaffold – scaffolding that is attached to the building via the entry of some of the poles into holes left in the brickwork by the bricklayer. The poles stay in position until the construction is complete and give the scaffold extra support.

No one other than a qualified **carded scaffolder** is allowed to erect or alter scaffolding. Although you are not allowed to erect or alter this type of scaffold, you must be sure it is safe before you work on it. You should ask yourself a number of questions to assess the condition and suitability of the scaffold before you use it:

- Are there any signs attached to the scaffold which state that it is incomplete or unsafe?

- Is the scaffold overloaded with materials such as bricks?

- Are the platforms cluttered with waste materials?

- Are there adequate guardrails and scaffold boards in place?

Chapter 3 Working at height

- Does the scaffold actually *look* safe?
- Is there the correct access to and from the scaffold?
- Are the various scaffold components in the correct place (see Figure 3.6)?
- Have the correct types of fittings been used (see Figure 3.7)?

> **Remember**
>
> If you have any doubts about the safety of scaffolding, report them. You could very well prevent serious injury or even someone's death

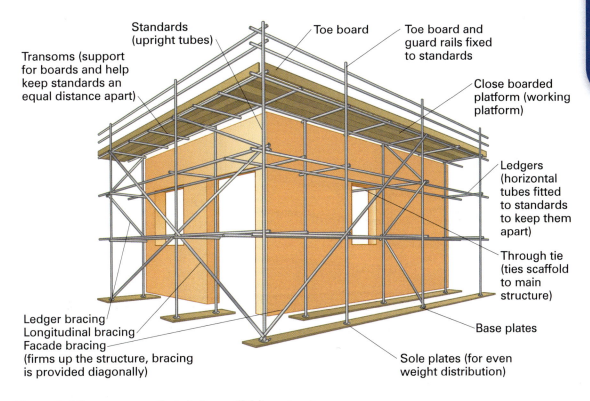

Figure 3.6 Components of a tubular scaffolding structure

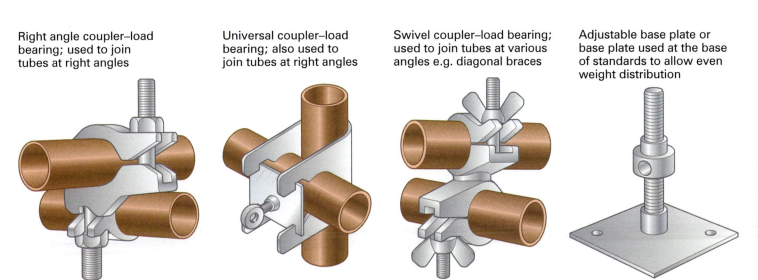

Figure 3.7 Types of scaffold fittings

79

Mobile tower scaffolds

Mobile tower scaffolds are so called because they can be moved around without being dismantled. Lockable wheels make this possible and they are used extensively throughout the construction industry by many different trades. A tower can be made from either traditional steel tubes and fittings or aluminium, which is lightweight and easy to move. The aluminium type of tower is normally specially designed and is referred to as a 'proprietary tower'.

Figure 3.8 Mobile tower scaffold

Low towers

These are a smaller version of the standard mobile tower scaffold and are designed specifically for use by one person. They have a recommended working height of no more than 2.5 m and a safe working load of 150 kg. They are lightweight and easily transported and stored.

Figure 3.9 Low tower scaffold

These towers require no assembly other than the locking into place of the platform and handrails. However, you still require training before you use one and you must ensure that the manufacturer's instructions are followed when setting up and working from this type of platform.

Erecting a tower scaffold

It is essential that tower scaffolds are situated on a firm and level base. The stability of any tower depends on the height in relation the size of the base:

- For use inside a building, the height should be no more than three-and-a-half times the smallest base length.

- For outside use, the height should be no more than three times the smallest base length.

The height of a tower can be increased providing the area of the base is increased **proportionately**. The base area can be increased by fitting outriggers to each corner of the tower.

For mobile towers, the wheels must be in the locked position whilst they are in use and unlocked only when they are being re-positioned.

There are several important points you should observe when working from a scaffold tower:

- Any working platform above 2 m high must be fitted with guardrails and toe boards. Guard rails may also be required at heights of less than 2 m if there is a risk of falling on to potential hazards below, i.e. reinforcing rods. Guardrails must be fitted at a minimum height of 950 mm.

- If guardrails and toe boards are needed, they must be positioned on all four sides of the platform.

- Any tower higher than 9 m must be secured to the structure.

- Towers must not exceed 12 m in height unless they have been specifically designed for that purpose.

- The working platform of any tower must be fully boarded and be at least 600 mm wide.

- If the working platform is to be used for materials then the minimum width must be 800 mm.

- All towers must have their own access and this should be by an internal ladder.

Definition

Proportionately – in proportion to the size of something else

Safety tip

Mobile towers must *only* be moved when they are free of people, tools and materials

Safety tip

Never climb a scaffold tower on the outside as this can cause it to tip over

Chapter 3 Working at height

FAQ

When using a ladder, is it OK to stand on the very top rung in order to reach something if I am quick? I'm only going to be a second.

You are putting yourself at risk if you do stand on the top rung and you would also be breaking health and safety law. It doesn't matter how quick you are, don't risk it and use a longer ladder or some scaffolding.

I haven't got enough trestles to support a scaffold board but it looks safe enough – is this OK?

If the scaffold board is not well supported by having trestles no further than 1.2 m apart, it could collapse or break. Make sure you have enough trestles.

On the job: Ryan's tower

Ryan is putting together a tower scaffold. He realises that he does not have enough diagonal supports for the tower so he decides to leave every other support out. Ryan is sure that this will be OK as the tower will still have some diagonal support. What do you think of Ryan's idea?

Knowledge check

1. Name four different methods of gaining height while working.
2. What must be done before any work at height is carried out?
3. What are your three health and safety duties when working at height?
4. As a rule, what is the maximum time you should work from a ladder or stepladder?
5. How should a wooden stepladder be checked before use?
6. When storing a wooden pole ladder, why does it need to be evenly supported along its length?
7. Explain the 1:4 (or 75°) ratio rule which should be observed when erecting a ladder.
8. When should a trestle platform be used?
9. What two types of board can be used as a platform with a trestle frame?
10. Why should you only use a specially designed hop-up?
11. There are two types of tubular scaffolding – what are they and how do they differ?
12. What are the eight questions you should ask yourself before using scaffolding?
13. In order to increase the height of a tower scaffold, what else has to be increased and by how much?
14. How high should scaffold guardrails be?
15. What is the only way you should access scaffolding?

chapter 4

Drawings

OVERVIEW

Drawings are the best way of communicating detailed and often complex information from the designer to all those concerned with a job or project. They are therefore one of the main methods of communication used in the building industry.

Drawings are part of the legal contract between client and contractor and mistakes, either in design or interpretation of the design, can be costly. Details relating to drawings must follow guidelines by the British Standards Institute: *BS 1192 Construction Drawing Practice*. This standardises drawings and allows everyone to understand them.

This chapter will help you to understand the basic principles involved in producing, using and reading drawings correctly.

The following topics will be covered:

- Types of drawing
- Drawing equipment
- Scales, symbols and abbreviations
- Datum points
- Types of projection
- Specifications.

Types of drawing

Working drawings

Working drawings are scale drawings showing plans, elevations, sections, details and location of a proposed construction. They can be classified as:

- location drawings
- component range drawings
- assembly or detail drawings.

Location drawings

Location drawings include block plans and site plans.

Block plans identify the proposed site by giving a bird's eye view of the site in relation to the surrounding area. An example is shown in Figure 4.1.

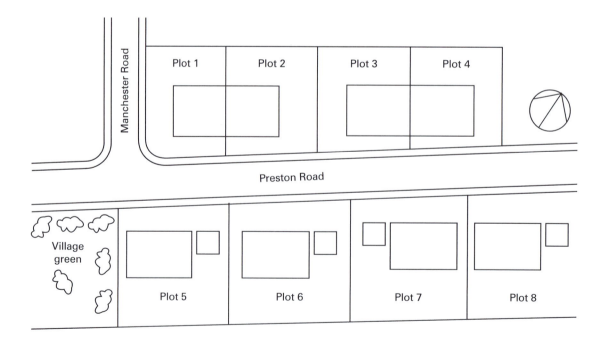

Figure 4.1 Block plan showing location

Site plans give the position of the proposed building and the general layout of the roads, services, drainage etc. on site. An example is shown in Figure 4.2.

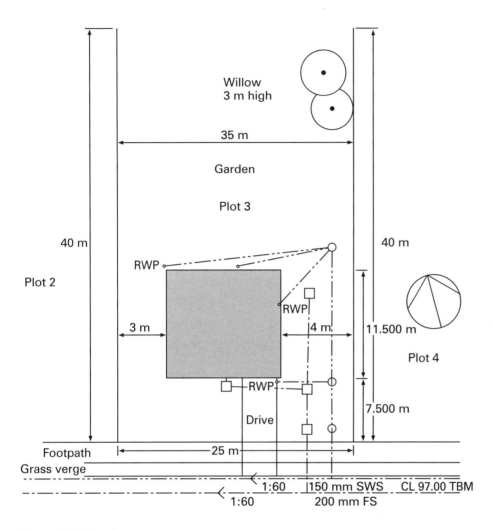

Figure 4.2 Site plan

Component range drawings

Component range drawings show the basic sizes and reference system of a standard range of components produced by a manufacturer. This helps in selecting components suitable for a task and available off-the-shelf. An example is shown in Figure 4.3.

Brickwork NVQ and Technical Certificate Level 2

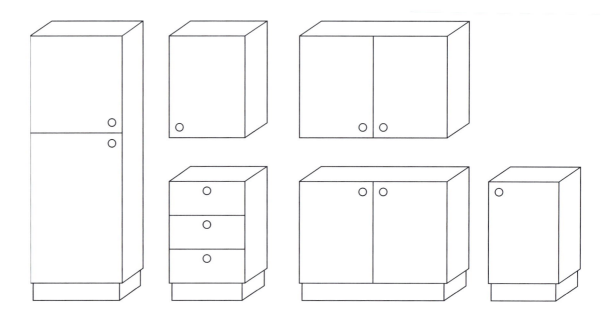

Figure 4.3 Component range drawing

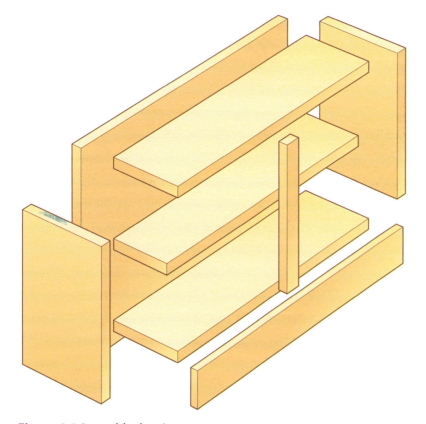

Figure 4.4 Assembly drawing

Assembly or detail drawings

Assembly or detail drawings give all the information required to manufacture a given component. They show how things are put together and what the finished item will look like. An example is shown in Figure 4.4.

Title panels

Every drawing must have a title panel, which is normally located in the bottom right-hand corner of each drawing sheet. See Figure 4.5 for an example. The information contained in the panel is relevant to that drawing only and contains such information as:

- drawing title
- scale used
- draughtsman's name
- drawing number/project number
- company name
- job/project title
- date of drawing
- revision notes
- projection type.

ARCHITECTS	CLIENT
Peterson, Thompson Associates 237 Cumberland Way Ipswich IP3 7FT Tel: 01234 567891 Fax: 09876 543210 Email: enquiries@pta.co.uk	Carillion Development
	JOB TITLE Appleford Drive Felixstowe 4 bed detached
	SCALE: 1:50
DRAWING TITLE Plan – garage	**DRAWING NO:** 2205-06
DATE: 27.08.2006	**DRAWN BY:** RW

Figure 4.5 Typical title panel

Remember

It is important to check the date of a drawing to make sure the most up-to-date version is being used, as revisions to drawings can be frequent

Drawing equipment

A good quality set of drawing equipment is required when producing drawings. It should include:

- set squares
- protractors
- compasses
- dividers
- scale rule
- pencils
- eraser
- drawing board
- tee square.

Brickwork NVQ and Technical Certificate Level 2

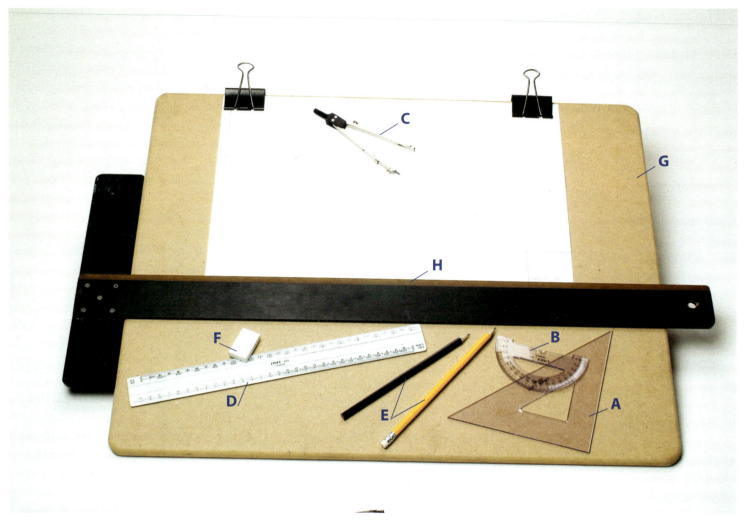

Drawing equipment

Set square

Two set squares are required, one a 45° set square and the other a 60° / 30° set square. These are used to draw vertical and inclined lines. A 45° set square (A) is shown in the photograph.

Protractor

Protractors (B) are used for setting out and measuring angles.

Compass and dividers

Compasses (C) are used to draw circles and arcs. Dividers (not shown) are used for transferring measurements and dividing lines.

Scale rules

A scale rule that contains the following scales is to be recommended:

1:5/1:50   1:10/1:100   1:20/1:200   1:250/1:2500

An example (D) is shown in the photo.

Pencils

Two pencils (E) are required:

- HB for printing and sketching
- 2H or 3H for drawing.

Eraser

A vinyl or rubber eraser (F) is required for alterations or corrections to pencil lines.

Drawing boards

Drawing boards (G) are made from a smooth flat surface material, with edges truly square and parallel.

T-square

The T-square (H) is used mainly for drawing horizontal lines.

Did you know?

Set squares, protractors and rules should be occasionally washed in warm soapy water

Scales, symbols and abbreviations

Scales in common use

In order to draw a building on a drawing sheet, the building must be reduced in size. This is called a scale drawing.

The preferred scales for use in building drawings are shown in Table 4.1.

Type of drawing	Scales
Block plans	1:2500, 1:1250
Site plans	1:500, 1:200
General location drawings	1: 200, 1:100, 1:50
Range drawings	1:100, 1:50, 1:20
Detail drawings	1:10, 1:5, 1:1
Assembly drawings	1:20, 1:10, 1:5

Table 4.1 Preferred scales for building drawings

These scales mean that, for example, on a block plan drawn to 1:2500, one millimetre on the plan would represent 2500 mm (or 2.5 m) on the actual building. Some other examples are:

- On a scale of 1:50, 10 mm represents 500 mm.

- On a scale of 1:100, 10 mm represents 1000 mm (1.0 m).

- On a scale of 1:200, 30 mm represents 6000 mm (6.0 m).

Why not try these for yourself?

- On a scale of 1:50, 40 mm represents:…………

- On a scale of 1:200, 70 mm represents:…………

- On a scale of 1:500, 40 mm represents:………….

The use of scales can be easily mastered with a little practice.

> **Remember**
>
> A scale is merely a convenient way of reducing a drawing in size

Variations caused through printing or copying will affect the accuracy of drawings. Hence, although measurements can be read from drawings using a rule with common scales marked, it is recommended that you work to written instructions and measurements wherever possible.

A rule marked with scales used in drawings or maps is illustrated in Figure 4.6. It should not be confused with a 'scale rule', using for measuring lengths, as shown in the photo on page 000, though often the same name is used.

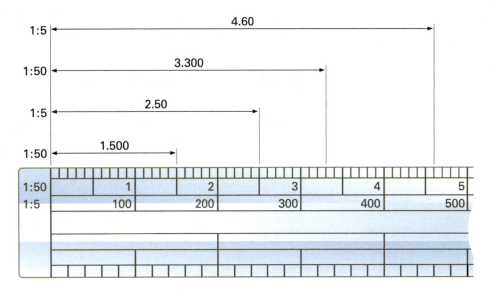

Figure 4.6 Rule with scales for maps and drawings

Symbols and abbreviations

The use of symbols and abbreviations in the building industry enables the maximum amount of information to be included on a drawing sheet in a clear way. Figure 4.7 shows some recommended drawing symbols for a range of building materials.

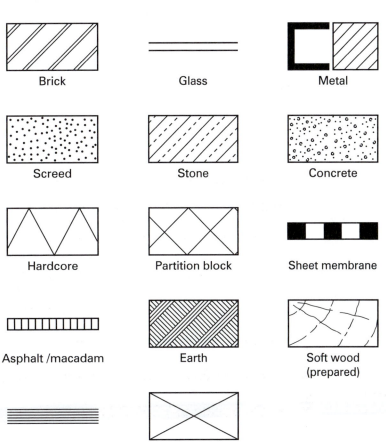

Figure 4.7 Building material symbols

Figure 4.8 shows some of the most frequently used graphical symbols, which are recommended in the British Standard BS 1192.

Figure 4.9 illustrates the recommended methods for indicating types of doors and windows and their direction of opening.

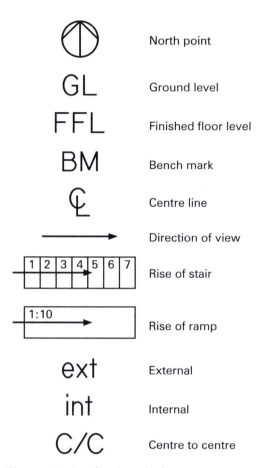

Figure 4.8 Graphical symbols

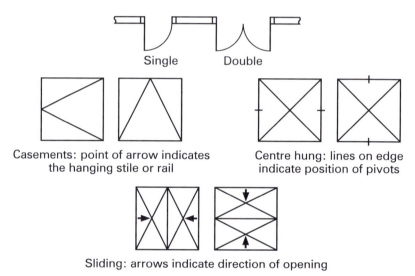

Figure 4.9 Doors and windows, type and direction of opening

Table 4.2 lists some standard abbreviations used on drawings.

Item	Abbreviation	Item	Abbreviation
Airbrick	AB	Cast iron	CI
Asbestos	abs	Cement	ct
Bitumen	bit	Column	col
Boarding	bdg	Concrete	conc
Brickwork	bwk	Cupboard	cpd
Building	bldg	Damp proof course	DPC

Table 4.2 Standard abbreviations used on drawings (contd opposite)

Item	Abbreviation	Item	Abbreviation
Damp proof membrane	DPM	Polyvinyl acetate	PVA
Drawing	dwg	Polyvinyl chloride	PVC
Foundation	fnd	Reinforced concrete	RC
Hardcore	hc	Satin chrome	SC
Hardboard	hdbd	Satin anodised aluminium	SAA
Hardwood	hwd	Softwood	swd
Insulation	insul	Stainless steel	SS
Joist	jst	Tongue and groove	T&G
Mild steel	MS	Wrought iron	WI
Plasterboard	pbd		

Table 4.2 Standard abbreviations used on drawings (contd)

FAQ

Why not just write the full words on a drawing?

This would take up too much space and clutter the drawing, making it difficult to read.

Datum points

The need to apply levels is required at the beginning of the construction process and continues right up to the completion of the building. The whole country is mapped in detail and the Ordnance Survey place datum points (bench marks) at suitable locations from which all other levels can be taken.

Ordnance bench mark (OBM)

OBMs are found cut into locations such as walls of churches or public buildings. The height of the OBM can be found on the relevant Ordnance Survey map or by contacting the local authority planning office. Figure 4.10 shows the normal symbol used.

Ordnance Survey Bench Mark (O.S.B.M.)

Figure 4.10 Ordnance Bench Mark

Site datum

It is necessary to have a reference point on site to which all levels can be related. This is known as the site datum. The site datum is usually positioned at a convenient height, such as finished floor level (FFL).

The site datum itself must be set in relation to some known point, preferably an OBM and must be positioned where it cannot be moved.

Figure 4.11 shows a site datum and OBM, illustrating the height relationship between them.

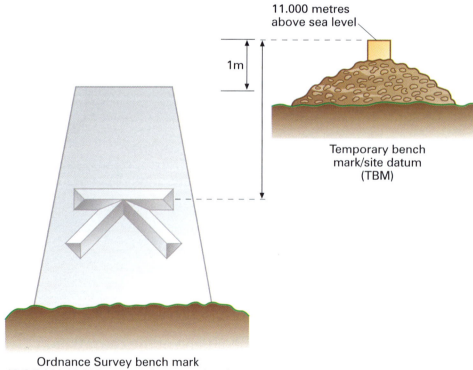

Ordnance Survey bench mark (O.S.B.M.) 10.000 metres above sea level

Figure 4.11 Site datum and OBM

If no suitable position can be found a datum peg may be used, its accurate height transferred by surveyors from an OBM, as with the site datum. It is normally a piece of timber or steel rod positioned accurately to the required level and then set in concrete. However, it must be adequately protected and is generally surrounded by a small fence for protection, as shown in Figure 4.12.

Temporary bench mark (TBM)

When an OBM cannot be conveniently found near a site it is usual for a temporary bench mark (TBM) to be set up at a height suitable for the site. Its accurate height is transferred by surveyors from the nearest convenient OBM.

All other site datum points can now be set up from this TBM using datum points, which are shown on the site drawings. Figure 4.13 shows datum points on drawings.

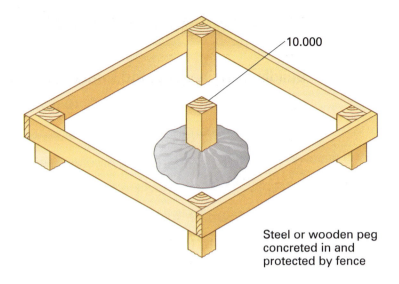

Figure 4.12 Datum peg suitably protected

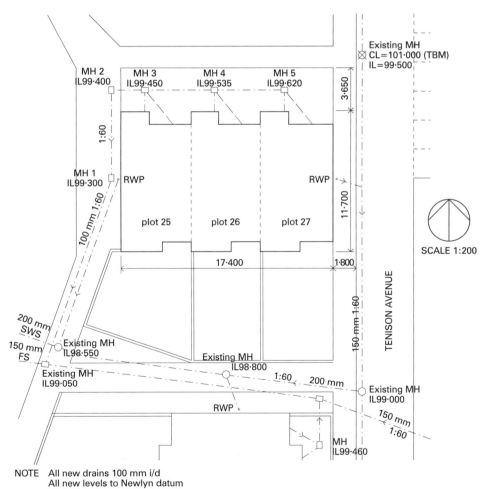

Figure 4.13 Example of datum points shown on a drawing

Types of projection

Building, engineering and similar drawings aim to give as much information as possible in a way that is easy to understand. They frequently combine several views on a single drawing.

These may be elevations (the view we would see if we stood in front or to the side of the finished building) or plan (the view we would have if we were looking down on it). The view we see depends on where we are looking from. There are then different ways of 'projecting' what we would see onto the drawings.

The two main methods of projection, used on standard building drawings, are orthographic and isometric.

Orthographic projection

Orthographic projection works as if parallel lines were drawn from every point on a model of the building on to a sheet of paper held up behind it (an elevation view), or laid out underneath it (plan view).

There are then different ways that we can display the views on a drawing. The method most commonly used in the building industry, for detailed construction drawings, is called 'third angle projection'. In this, the front elevation is roughly central. The plan view is drawn directly below the front elevation and all other elevations are drawn in line with the front elevation. An example is shown in Figure 4.14.

Chapter 4 Drawings

Figure 4.14 Orthographic projection

Isometric projection

In isometric views, the object is drawn at an angle where one corner of the object is closest to the viewer. Vertical lines remain vertical but horizontal lines are drawn at an angle of 30° to the horizontal. This can be seen in Figure 4.15, which shows a simple rectangular box.

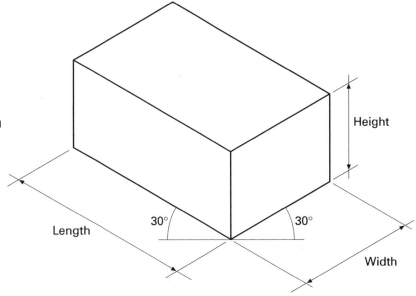

Figure 4.15 Isometric projection of rectangular box

99

Figures 4.16 and 4.17 show the method of drawing these using tee square and set squares.

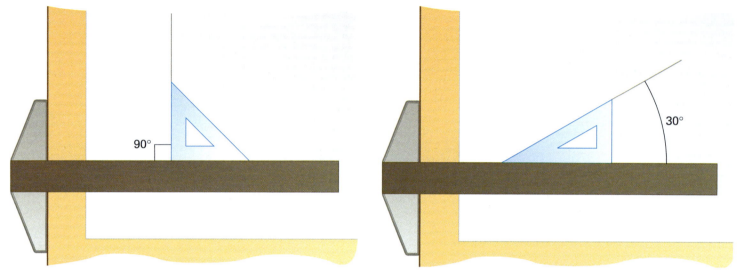

Figure 4.16 Drawing vertical lines

Figure 4.17 Drawing horizontal lines

Specifications

Except in the case of very small building works, drawings cannot contain all of the information required by the contractor; in particular, the standard of materials to be used and quality of workmanship. For this purpose the architect will prepare a document called a specification to supplement the working drawings.

The specification is a precise description of all the essential information and job requirements that will affect the price of the work, but cannot be shown on drawings.

Typical items shown on the specification are:

- site description
- restrictions (limited access, working hours etc.)
- services and availability of services (waste, gas, electricity, telephone etc.)
- workmanship (quality, size tolerances, finishing requirements etc.)
- other information (nominated suppliers, sub-contractors, site clearance etc.)

On the job: Setting out using drawings

Jason is using a drawing to set out the internal walls for a new house. He has the overall internal measurements, from left-side cavity wall to right-side cavity wall. The measurements for the room sizes are on the drawings along with the thickness of the dividing walls. When Jason adds together the room sizes and wall thickness, he gets a total that is 170 mm more than the overall internal measurement on the drawings. What should Jason do?

Knowledge check

1. Briefly explain why drawings are used in the construction industry.

2. What do the following abbreviations stand for: DPC; hwd; fnd; DPM?

3. Sketch the graphical symbols which represent the following: brickwork; metal; sawn timber; hardcore.

4. Can you name the main types of projection, which are used in building drawings?

5. What does a block plan show?

6. What are dividers used for?

7. What type of information could be found in a drawing's title panel?

8. Name two ways in which you can find out the height of a OBM?

9. In isometric projection, at what angle are horizontal lines drawn?

10. What type of information can be found in specifications?

chapter 5

Hand tools and equipment

OVERVIEW

This chapter looks at the many tools and different types of equipment used by a bricklayer. If a bricklayer is working on a new housing site, he will use fewer tools than a bricklayer who carries out extensions and alteration work. Most bricklayers carry a wide selection of tools so as not to be caught out if there are changes to the type of work to be undertaken or if extra work is required.

Most bricklayers collect their tools over a long period of time, usually adding tools as they require them, normally to carry out a specific job at that time or sometimes to replace worn out tools. Larger tools they may use, such as drills, breakers and cutters may be bought or hired, depending on how often the tool is used and the cost to buy compared with hire charges.

We will cover the following topics in this chapter:

- Health and safety
- Hand tools
- Portable power tools.

Health and safety

All tools are potentially dangerous and you, as an operator of tools, must make sure that all health and safety requirements relating to the tools are always carried out. This will help ensure that you do not cause injury to yourself and, equally important, to others who may be working around you and the general public. Make sure you follow any instructions and demonstration you are given on the use of tools, as well as any manufacturers' instructions provided on purchase of the tool.

Basic health and safety rules

1. Always make sure you use the correct personal protective equipment (PPE) required to use the tool and do the job you are carrying out.

2. Never 'make do' with tools. Using the wrong tool for the job usually breaks health and safety laws.

3. Never play or mess around with a tool regardless of the type, whether it is a hand tool or power tool.

4. Never use a tool you have not been trained to use, especially a power tool.

Hand tools

The bricklayer's tools are his living. Therefore great care should be taken when using tools. Always make sure they are correctly cleaned after use. When purchasing tools, make sure they are of good quality, durable and the right tool for the job.

The tools covered in this chapter are a large selection of those that you will come across during your early years in the trade, depending on the type of work carried out during this time.

> **Remember**
>
> Follow the basic health and safety rules and you (and others) will be safe

> **Remember**
>
> It does not always pay to buy the cheapest tool available. However, your finances may dictate the quality you can afford to buy during the early years of training, so you may find you need to replace tools when you can afford to

Chapter 5 Hand tools and equipment

Trowels

Brick trowel

The brick trowel is used by the bricklayer to take the mortar off the mortar board, lay it on to the wall, and spread it to form a uniform bed joint ready for the bricks to be laid upon.

Brick trowels can be purchased in different sizes and for left- or right-handed people. They are made from solid rolled forged carbon steel, with a hickory handle. Always clean your trowel with water after use, dry it thoroughly and lightly oil to prevent rust from forming.

Brick trowel

Pointing trowel

Pointing trowels can be purchased in different sizes depending on requirement and preference. They are used for filling in joints and pointing certain types of joint finishes. They are made from solid rolled forged carbon steel, with a hickory handle. Clean as you would a brick trowel.

Pointing trowel

Concrete or plastering trowel

This trowel is sometimes called a float and is used for finishing concrete, for putting plaster on to walls and ceilings and rendering on to walls. It is often made of steel with a wood or plastic handle and comes in different sizes. Some are made of plastic or wood instead of steel, which means they are lighter if using for long periods of time, especially in plastering. They are also used for floor **screeding** finishes. Clean as you would a brick trowel.

Concrete or plastering trowel (plasterer's float)

105

Hand hawk

Hand hawk

The hand hawk is used by the bricklayer in the process of filling joints in pointing and jointing. The mortar is placed on the top and used with the pointing trowel to avoid having to repeatedly return to the mortar board. The hawk is usually made of wood or plastic but some plasterers use steel hawks. Clean it with water after use.

Joint raker

Joint raker (chariot)

The joint raker often goes under the name 'chariot'. This tool is used for raking out soft mortar joints, whether to give a joint finish on new work or to take out old mortar ready for repointing. Made of steel, it has an adjustable raking pin to change the depth the joint is recessed.

Jointer

Jointer

Usually made from steel, this tool is used to form the joint finish after building a wall. This serves two purposes:

1. It helps to make the joint more waterproof.
2. The appearance of the wall looks more pleasing.

The tool shown in the photo is used to form a half rounded joint referred to as a **bucket handle finish**. Jointers are made from rolled forged carbon steel.

Chapter 5 Hand tools and equipment

Tape measure

Tape measures come in many sizes, from 3 m up to 10 m for the pocket type, and from 10 m up to 30 m for larger setting out tapes – some can go up to 100 m. Used for measuring or checking sizes, they are usually made with plastic or steel cases, with a steel measure. Some come with both metric (centimetres and metres) and imperial (inches and feet) measurements, although most only have metric measures. Larger tapes are made of steel or fibreglass.

Steel tape measure

Always ensure tapes are wiped clean after use as moisture will result in the steel rusting and any dirt will clog up the tape, making it hard to release or retract.

A 30 m tape measure

Did you know?

The use of a lightly oiled rag when retracting a tape measure helps stop rusting

Screwdrivers

There are many different types of screwdriver but a straightforward slotted and Phillips-type screwdriver are always used by bricklayers for fixings etc. Screwdrivers are made from steel with a plastic handle grip.

Slotted and Phillips screwdrivers

107

Hammers

Brick hammer

Brick hammers are used for cutting and shaping bricks and are made from forged steel with a hickory handle. The brick should be held stable, with the hammer held in the appropriate hand. With the square edge of the hammer you tap the brick at the position you require the cut.

Goggles and mask should be used to prevent chippings going into your eyes and dust being inhaled. Also, carry out cutting in a separate place away from other people to prevent them from being affected by your actions.

Safety tip

Great care must be taken when using a brick hammer as it is very easy to hit your supporting hand with the hammer

Brick hammer

Club/lump hammer

Club or lump hammer

This is a heavy hammer used together with a bolster chisel for cutting bricks/blocks by hand. It is also used with other chisels for cutting out bricks, knocking holes through walls and removing joints using a plugging chisel. It is made from forged steel, with a hickory handle, and comes in different weights, usually ranging from 2.5 – 4 lb.

Chapter 5 Hand tools and equipment

Claw hammer

The claw hammer is used for fixing nails into different materials or taking out existing nails with its claw. It should not be used in conjunction with chisels as the rounded head is small and the hand is usually an easy target to hit. It is made from forged tempered steel with a hardened head and a polished finish.

Claw hammers come in different weights ranging from 16 oz to 24 oz.

Claw hammer

Did you know?

The claw hammer can come with a wooden or steel shaft, the steel shaft normally having a rubber shaft cover to absorb the shock

Scutch or comb hammer

This is used to trim bricks or blocks to the correct size or shape. It is made from forged tempered steel, with a hickory handle. Small combs are inserted into the head to give the trimming blade and it can come with a single or double scutch head.

Great care must be taken when using this hammer as particles of the materials being trimmed fly off. Also, you must be careful not to hit your hands.

Scutch or comb hammer

Remember

Change the combs regularly as blunt ones make the work take longer

Safety tip

Always use suitable personal protective equipment (PPE)

109

Chisels

There are many different types of chisel used by a bricklayer. Some of the most frequently used are described below.

Bolster chisel

This is used mainly for cutting bricks or blocks to the required size and angled shape. It is made from hardened tempered steel, usually in cutting blade sizes from 64 mm to 100 mm. Some come with safety handle grips.

Bolster chisels

Cold chisels

Cold chisels

These are used for cutting holes or knocking down brickwork/blockwork during alteration work. They are made from forged hardened tempered steel and come in a variety of different sizes ranging from 6.5 mm to 25 mm blade cutting size. Some come with safety handle grips. Different shaft lengths are available, ranging from 100 mm in length up to 450 mm.

Chapter 5 Hand tools and equipment

Plugging chisel

This type of chisel is used for taking out existing mortar joints for repairs and cutting out existing joints for lead work. It is made from fluted cast steel and is normally available in just one size.

Plugging chisel

Scutch comb chisel

Used for trimming bricks or cutting electrical chases into walls, it holds a trimming comb at the end and is made of hardened steel. The trimming blade size is 25 mm.

Scutch comb chisel

> **Remember**
>
> Make sure you change the comb of a scutch chisel on a regular basis as blunt ones take more time and effort to use

Maintenance

Care must be taken when using chisels to make sure the blades are always sharp. Also, make sure that the striking end does not form a 'mushroom' through hitting, as parts of this are likely to break off. This could result in injury to the user's face or, if parts break off when holding, they could penetrate the skin on the hand. If 'mushrooming' does occur, the end must be ground smooth before further use.

Spirit levels

Made from aluminium, spirit levels come in various sizes from 225 mm to 1200 mm, with 1200 mm being the main size that a bricklayer uses. They are used for levelling things horizontally and for plumbing vertically, having bubbles that give a reading between set lines to determine accuracy of the work. Some levels have an adjustable bubble at the bottom for levelling angled work.

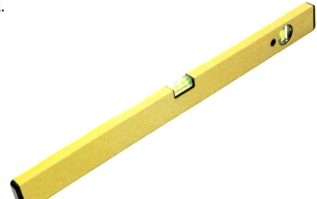

Spirit level

Definition

True – giving an accurate reading

Great care must be taken when using levels as they can easily go out of **true**, which can result in work seeming to be level or plumb by the reading but actually being wrong. This could result in work having to be taken down and redone. The main cause of a spirit level going out of true is normally that the level has been hit with a trowel or hammer, dropped or in other ways misused. Always wash the spirit level off with water after use to keep it clean.

Boat level

Boat levels

These are useful for levelling and plumbing small work such as soldier bricks and are always handy to have in your toolkit. They work on the same principle as the larger levels.

Checking a level for accuracy (true)

Before you use a level, it is best to check it for accuracy (true) first. You can do this by simply placing the level on a flat surface or on previously levelled screws, making a mark at both ends and reading the bubble position. You should then turn the level around, reversing the positions of the ends and make sure they touch your marks. The bubble can then be read again. If the bubble is in the same position both times you read it, the level is accurate. If the bubbles give different readings, the level is out of true and the bubble will need adjusting to make the level accurate. See Figure 5.1.

Did you know?

You can use the same method to check a level for true upright against a surface (i.e when using to plumb)

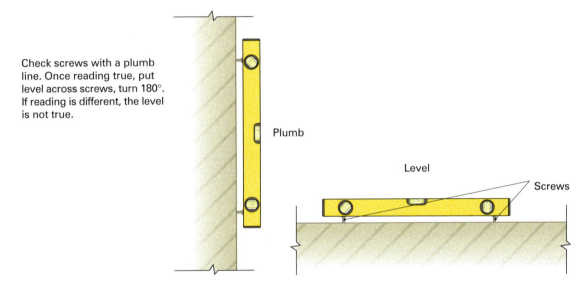

Figure 5.1 Checking levels for accuracy

Lines and pins

Lines and pins are used for laying bricks and blocks once the corners have been erected to ensure work is in a straight line. The pins are placed into the **perp** joints at each end of the run so that the line runs from the top of the laid brick at each end. This ensures that the bricks that are to be laid run in a straight line and, by putting the top of the brick to the line, also keeps the bricks level between the two points, assuming that the corners are correct.

Definition

Perp – the vertical joint between two bricks or blocks

The pins are normally made from forged steel for light duty work. Some are made from thicker steel for heavier duty work. The line can be made of nylon or cotton. Nylon is more durable but less flexible, whereas cotton is the opposite. While using either, care should be taken when laying the bricks as they cut very easily.

Remember

The line comes in different lengths so make sure it is long enough for the work to be undertaken

Brick lines and line pins

Corner blocks

Corner blocks are used to attach the line to keep the brickwork straight. They are made from wood, plastic or steel and fit on to the corners of the brickwork with the lines pulled tight to hold them in place. They are then raised to complete each course as it progresses.

Corner blocks

Chapter 5 Hand tools and equipment

Tingle plate

This is used to stop the line from sagging in the middle on longer runs of brickwork. The central brick of the wall should be set to the correct gauge and the tingle plate set on top with the line passing through to stop sagging. A tingle plate is made of steel sheet and is worth having in your kit. It could stop work from having to be rebuilt due to the wall sagging in the middle.

Corner profiles

Corner profiles are used to save time, as there is no need to build corners. They are secured to existing brickwork at the bottom and are adjustable so as to allow for making plumb. They are made of right-angled steel and the bricks sit hard against the internal corner of the profile when building. If set up correctly, once removed after completion of the brickwork, the corner should be plumb on both edges, which should be more accurate than hand-built corners using the conventional method. Profiles can be set on each corner. They can then be marked for correct gauge and lines pulled from corner to corner. This speeds up the building process.

Did you know?

Corner profiles are quite expensive to buy but pay for themselves as bricks can be laid faster

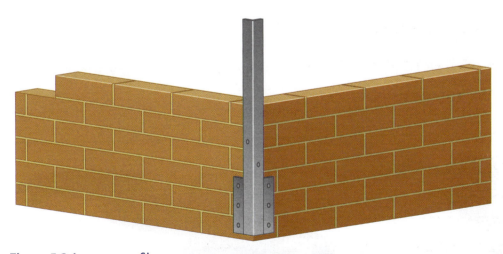

Figure 5.2 A corner profile

115

Portable power tools

Portable power tools come in different forms: battery powered, electric and petrol powered. You should not use any power tools unless you have been fully trained in how to do so. Most power tools used by a bricklayer are for cutting, breaking or drilling.

As with most tools, PPE must always be worn. On site, the type of tool used may depend on the circumstances – if no electricity is available, only battery or petrol power could be used.

Angle grinders

Angle grinders are cutting tools that run by electricity using 110 V and 230 V supplies or are battery powered. They cut using an abrasive or diamond-type disc. They range from 100 mm to 230 mm diameter disc size.

Angle grinders are used by bricklayers mainly for cutting bricks, blocks, concrete and stone to size or for cutting existing material for alteration. Great care should be taken when using them as the disc travels at very high speed and takes time to slow down after release of the trigger, but can still cut.

Owing to the cutting speed, large amounts of dust and particles are released from the material, so goggles and mask should always be worn in addition to the usual PPE. Ear defenders should also be worn. All leads should be checked before and after use for cuts or splits, and with a 110 V supply a transformer must be used.

Did you know?

In the construction industry, only 110 V type or battery type angle grinders are allowed to be used on site by health and safety law

Safety tip

No person is allowed to change a cutting disc unless they hold an Abrasive Wheels Certificate

100 mm/4" electric grinder

225 mm/9" electric grinder

Chapter 5 Hand tools and equipment

Petrol cutters

This type of cutter is the same as the angle grinder but is motorised, running on petrol. The disc size is 300 mm and it runs at a faster speed than the angle grinder. It uses abrasive or diamond cutting discs and is used for heavier duty cutting. Great care must be taken when using, and fuel and oil levels should be checked regularly. Discs should be also be regularly checked for wear or damage. To ensure that no movement occurs when the disc is spinning, the locking nut is securely tightened and the shim (a flat metal plate) is fitted over the main spindle either side of the disc.

12" Petrol cutter

On the job: Using a disc cutter

Josh is asked by his supervisor to cut 30 bricks to a set size using a petrol cutter. He has had training and is competent using it but he is not sure what blade is on the cutter as it is covered up by the shim that holds it tight. What should he do? What type of PPE should Josh be wearing?

Did you know?

You have to be 18 or over to use a disc cutter on site

Breakers

These tools are used for heavy duty work like breaking up road surfaces, concrete floors, foundations and brickwork. They run on electricity or by compressed air. Breakers come in various sizes to suit the type and density of the material to be broken up. They can be fitted with points and chisels for the breaking operation, which uses a hammer vibration technique to break up materials.

Bricklayers are more likely to use 110 V electric breakers than compressed air breakers, which are used more for road surface works. In addition, compressed air breakers are harder to move around on site due to the bulky compression unit required.

In addition to the normal PPE, ear defenders should always be worn, as well as goggles to protect the eyes from particles and a mask for protection from dust. PPE should be of the correct standard and capable of doing its job. Faulty PPE should not be used and must be reported to your supervisor and thrown away.

Safety tip

Breakers should only be used for short periods of time due to the vibration effect on the hands and body

On the job: Using a breaker

Femi is on his second day on site when his boss asks him to use a breaker to break up an area of concrete. His boss tells him the breaker is in the shed and he is to help himself. The breaker is a brand new one in a box with the instruction manual inside explaining how to use it. Femi has never used one before. What should he do?

Breaker

Chapter 5 Hand tools and equipment

Drills

Drills can be electric or battery powered and come in many different sizes. They are used for drilling and fixing different types of components, having many types of drill bits suitable for different jobs.

Electric drills vary from 500 W to 1150 W, with battery drills ranging from 7.2 V to 24 V. Drill bits for masonry drilling range from 4 mm up to 25 mm and come in a variety of different lengths, with bits available for wood and metal drilling.

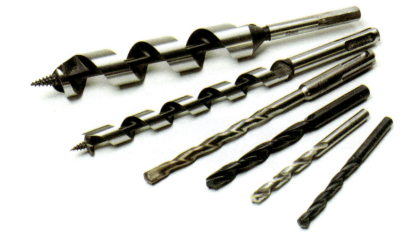

Electric drill and various bits

Scabblers

This is a type of drill with percussion movement and is used for chiselling, chasing and removal of existing plaster and rendering. Different types of bits can be used such as points, chisels and a comb alternative.

Scabbler

119

Block cutters

These are used for cutting bricks, blocks, paviers and concrete slabs and have a sharp cutting edge and pressured compression to break the material to the correct size. They can be moved around fairly easily but are more time-consuming than cutting bricks and blocks by hammer and bolster. However, they are better for slabs and paviers, rather than cutting with disc cutters or grinders.

Block cutter

Knowledge check

1. What is a scutch hammer used for?
2. What do you need to change a blade on a disc cutter?
3. What does PPE stand for?
4. What is a tingle plate used for?
5. To use any power tool what must you have had?
6. What is a hawk used for?
7. What extra PPE should be worn when using a disc cutter?
8. What is a bolster used for?
9. How do you check a level for accuracy?
10. What is another name for a joint raker?

chapter 6

Handling and storage of materials

OVERVIEW

This chapter is based on the moving and storage of materials once they have been received on to the building site. You will cover these areas during both NVQ Level 1 and Level 2 of your qualification, learning how to stack and store materials safely, securely and without risk of injury. Once materials have been offloaded, it is the responsibility of the building contractor to move the materials to a safe and secure place until they are required for use.

This chapter covers:

- Moving materials
- Delivery of materials
- Storage of materials.

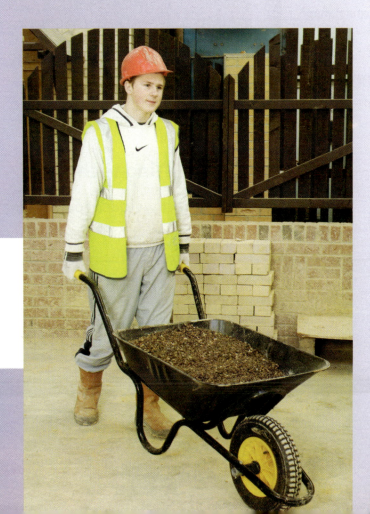

121

Moving materials

Lifting heavy or awkward objects, such as bags of cement, can cause injury if not performed correctly. Incorrect lifting techniques put extra stress on to the lower back. After years of 'bad' lifting, the discs between the **vertebrae** become disjointed and could slip. This might even happen when you are lifting a relatively light load.

Definition

Vertebrae – the small bones that form the backbone

Did you know?

When lifting the 'correct' way, you will be using the stronger leg muscles rather than the weak back muscles, you will keep the natural shape of the spine, and not pull it apart

Incorrect method for lifting – the legs are straight and the lift is coming from the back, instead of the legs

Kinetic lifting technique

To lift correctly, crouch to the object with your feet slightly apart with one foot at the side of the object, facing the direction in which you wish to go. Keep your back straight at all times, take a firm grip with both hands and lift by standing up.

Safe lifting technique

Chapter 6 Handling and storage of materials

To enable a person to take a firm grip, heavy or awkward objects should be stacked on timber bearers, to enable the person lifting to get their fingers right under the load. Also, when lowering the object it makes sense to again place timber bearers below to prevent you from getting trapped fingers.

If an object has been assessed as being too heavy or awkward for one person to lift, then a work colleague should help. When team lifting, ideally everyone should be about the same height and build, but this is not always possible as everyone is different. The effort should be the same for each person, and only one person should give instructions. The instructions should be given clearly, using a recognised call like 'lift on three: one, two, three'.

For more information on safe manual handling, see Chapter 2 Health and safety (page 33).

Before lifting any object, you must wear suitable protective clothing, such as boots, gloves, helmet and overalls when necessary. Examine the load to test if you can lift it; if not get help. Check the route over which you are about to take the object: check there are no obstructions and that the area where you are about to place it is clear.

Carrying cement can be very tiring, as the bags are very awkward. The easiest way to carry bagged items is on the shoulder.

> **Safety tip**
>
> Never be afraid to admit something is too heavy for you to lift. If you don't, you may end up injuring yourself

> **Safety tip**
>
> It may make it easier to carry bags of cement if you support the shoulder by putting your hand on your hip, and hold the bag with your other hand (see photo)

Correct method for carrying a bag of cement

123

Remember

When lifting materials over long distances, use a wheelbarrow

Loose materials such as aggregates, bricks and bags of cement may be transported by using a wheelbarrow. Building sites are usually quite rough terrain, and it is hard work pushing loaded barrows over rough ground. Barrow 'runs' may be used, which are boards laid down over the rough ground, but if runs are over trenches, make sure the boards are thick enough to carry the weight. Load the barrow evenly with most of the weight over the wheel.

Remember

When moving any materials, care must be taken not to damage the material or the packaging. Do not throw bricks into a wheelbarrow; place them in neatly. When unloading they should be unloaded by hand, not tipped

On the job: Moving a concrete lintel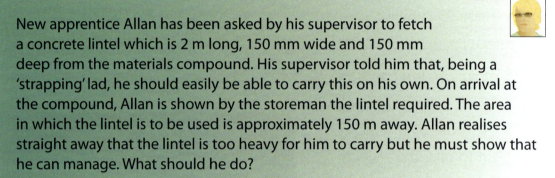

New apprentice Allan has been asked by his supervisor to fetch a concrete lintel which is 2 m long, 150 mm wide and 150 mm deep from the materials compound. His supervisor told him that, being a 'strapping' lad, he should easily be able to carry this on his own. On arrival at the compound, Allan is shown by the storeman the lintel required. The area in which the lintel is to be used is approximately 150 m away. Allan realises straight away that the lintel is too heavy for him to carry but he must show that he can manage. What should he do?

Delivery of materials

When materials are delivered on to a building site, someone must sign a delivery note. This note is evidence that a delivery has been made, and that all the items on that note have been received in good condition. Before the delivery note is signed, the delivery must be checked, ensuring that all the items are in good condition and counted. If there is any **discrepancy** in the delivery note and the materials, the note must not be signed, and the supervisor or foreman informed.

Definition

Discrepancy – a difference between what there is and what there should be

If a supervisor is not available and a shortage is noted in the delivery, the person signing should write on the note what is missing, and then sign the copy. Figure 6.1 shows a typical delivery note.

Chapter 6 Handling and storage of materials

Delivery note

Bailey & Sons Ltd

Building materials supplier

Tel: 01234 567890

Your ref: AB00671

Our ref: CT020

Date: 17 Jul 2006

Order no: 67440387

Invoice address:
Carillion Training Centre,
Deptford Terrace, Sunderland

Delivery address:
Same as invoice

Description of goods	Quantity	Catalogue no.
OPC 25kg	10	OPC1.1

Comments:

Date and time of receiving goods:

Name of recipient (caps):

Signature:

Figure 6.1 A typical delivery note

Did you know?

If the delivery note was signed without checking the contents of the delivery, it could result in the company, or even the person who signed, paying for materials that were not delivered

Remember

Make sure you pass the delivery note to your supervisor as soon as possible after receiving delivery

Storage of materials

Aggregates

Aggregates are granules or particles that are mixed with cement and water to make mortar and concrete. Aggregates should be hard, durable and should not contain any form of plant life, or anything that can be dissolved in water.

Aggregates are classed in two groups:

1. Fine aggregates are granules that pass through a 5 mm sieve.
2. Coarse aggregates are particles that are retained by a 5 mm sieve.

The most commonly used fine aggregate is sand. Sand may be dug from pits and riverbeds, or dredged from the sea.

Did you know?

Sea sand contains salts, which may affect the quality of a mix. It should not be used unless it has been washed and supplied by a reputable company

Remember

Poorly graded sands, with single size particles, contain a greater volume of air and require cement to fill the spaces

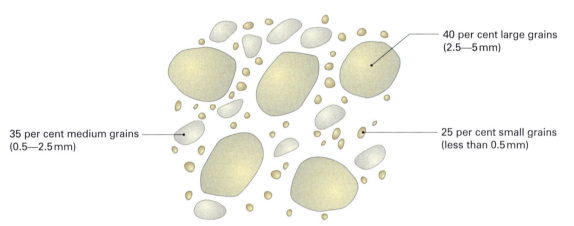

Figure 6.2 Sand particles

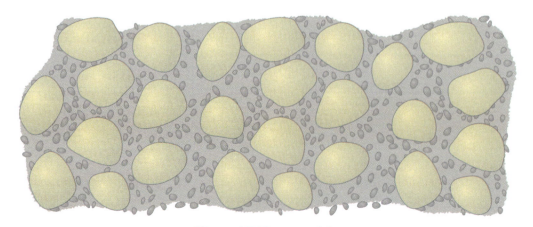

Figure 6.3 Mortar particles

Good mortar should be mixed using 'soft' or 'building' sand. It should be well graded, which means having an equal quantity of fine, medium and large grains.

Chapter 6 Handling and storage of materials

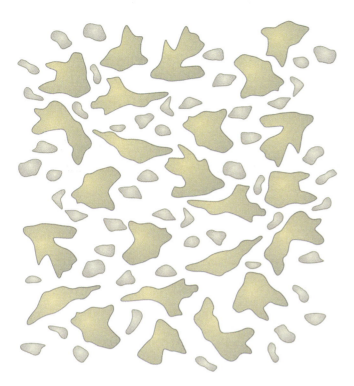

Figure 6.4 Concrete particles

Concrete should be made using 'sharp' sand, which is more angular and has a coarser feel than soft sand, which has more rounded grains.

When concreting, we also need 'coarse aggregate'. The most common coarse aggregate is usually limestone chippings, which are quarried and crushed to graded sizes, 10 mm, 20 mm or even larger.

Remember

The different sizes of aggregates should be stored separately to prevent the different aggregates getting mixed together

Storage of aggregates

Aggregates are normally delivered in tipper lorries, although nowadays one-tonne bags are available and may be crane handled. The aggregates should be stored on a concrete base, with a fall to allow for any water to drain away.

In order to protect aggregates from becoming contaminated with leaves and litter it is a good idea to situate stores away from trees and cover aggregates with tarpaulin or plastic sheets.

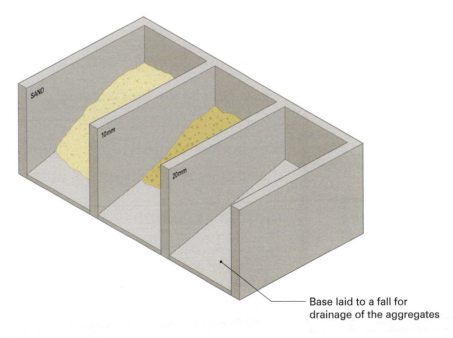

Figure 6.5 Bays for aggregates

127

Brickwork NVQ and Technical Certificate Level 2

Did you know?

Hardcore must be clean and not contain any water-soluble sulphates, which may damage the concrete

Remember

Always use the correct cement for the job – never make do

Did you know?

In order for cement to be sold, it must pass a series of tests for fineness, cleanliness, strength and setting times. These tests are controlled by the British Standards Institution who then award a BS Kite Mark and number

Hardcore

Hardcore is a material that is usually placed below concrete floors and pathways to provide a hard base on which to lay concrete. It is made up of broken bricks, clean clinker, stone, recycled crushed concrete or any similar inert material. All hardcore must be smaller than 100 mm.

Cement

Cement is a material which when mixed with water goes through a chemical reaction, **hydration**, which causes the cement powder to turn very hard. Cement is made from chalk or limestone and clay, which is crushed into a powder and mixed together and heated in a rotary kiln before being ground into a fine powder. This is then bagged and distributed to builder's merchants.

There are many different types of cement for various situations.

Ordinary Portland Cement (OPC)

For use in concrete, mortar or grout. When hardened it looks like Portland stone.

Rapid-hardening Portland cement

This is not like the 'quick-drying' cement used in cartoons! It does not go hard in seconds, just in less time than Ordinary Portland Cement. It is used where high, early strength is required. Concrete that is mixed with this cement has the advantage of being put in use more quickly than if mixed with OPC.

Masonry cement

This is a Portland cement that has a plasticising and air entraining material added.

Sulphate-resisting cement

Used for concrete or mortars exposed to sulphate attack, i.e. below ground level or in fireplace construction.

Chapter 6 Handling and storage of materials

Plaster

Plaster is made from gypsum, water and cement or lime. Aggregates can also be added depending on the finish desired. Plaster provides a joint-less, smooth, easily-decorated surface for internal walls and ceilings.

Gypsum plaster

This is for internal use and contains different grades of gypsum. Plasters are available depending on the background finish. Browning is usually used as an undercoat on brickwork but in most cases, a one-coat plaster is used, and on plasterboard, board finish is used.

Cement-sand plaster

This is used for external rendering, internal undercoats and waterproofing finishing coats.

Lime-sand plaster

This is mostly used as an undercoat, but may sometimes be used as a finishing coat.

Storage of cement and plaster

Both cement and plaster are usually available in 25 kg bags. The bags are made from multi-wall layers of paper with a polythene liner. Care must be taken not to puncture the bags before use. Each bag, if offloaded manually, should be stored in a ventilated, waterproof shed, on a dry floor on pallets. If offloaded by crane they should be transferred to the shed and the same storage method used.

The bags should be kept clear of the walls, and piled no higher than five bags. It is most important that the bags are used in the same order as they were delivered. This minimises the length of time that the bags are in storage, preventing the contents from setting in the bags, which would require extra materials and cause added cost to the company.

Did you know?

On larger sites some companies use a machine spray system to cover large areas with plaster quickly, using many plasterers to complete the work

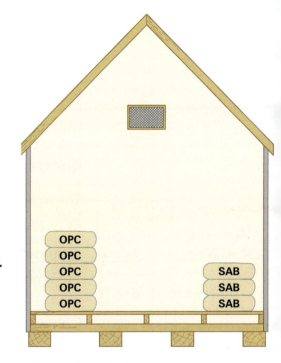

- Dry, ventilated shed
- Stock must be rotated so that old stock is used before new
- Not more than five bags high
- Clear of walls
- Off floor

Figure 6.6 Storage of cement and plaster bags in a shed

129

Levelling compounds

To level uneven floors before laying final covering, a self-levelling compound should be used, which creates a smooth surface and should be covered within 24 hours of applying. Compounds are available in bags of 10, 12.5, 15 and 25 kg. Storage is the same as for plaster and cement.

Bricks

Manufacture of bricks

Bricks are rectangular units used to build walls etc. They are normally made from clay, concrete or calcium silicate.

Clay bricks

Clay bricks are usually pressed, cut or moulded and then fired in a kiln at a very high temperature. Their density, strength, colour and surface texture will depend on the variety of clay used and the temperature of the kiln during firing.

Concrete bricks

These are made from a mixture of Portland cement and aggregate of sand or crushed rock, with a colouring agent added if required. They are made by thumping, vibrating or pressing, and will develop their strength naturally or by steam curing.

Calcium silicate (sand-lime) bricks

These are made from sand and lime or crushed flint and lime, pressed into shape and steamed at a high temperature in an **autoclave**. Pigments may be added during the manufacturing process to achieve a range of colours.

Types of brick

Bricks are classified according to the type of wall they can be used to build.

> **Definition**
>
> **Autoclave** – a machine that uses high pressure and temperature to cause chemical reactions

Common bricks

These have no particular finish to the surface, and are generally used for internal walls or for any work that is to be covered.

Facing bricks

These have a finished surface, sanded smooth or textured. They may be uniform in colour or multi-coloured. They are used to provide a long-lasting and pleasing finish to buildings.

Engineering bricks

These are very dense bricks used where strength is the most important quality of the building. They are used to build inspection chambers, retaining walls and piers carrying heavy loads, bridge abutments, steelwork and to line concrete chimney shafts.

How bricks are identified

Bricks are often identified by manufacturers' names. These are taken from:

1. their place of origin (e.g. Swanage, Ibstock, Leicester)
2. their colour (red, buff, blue, multi-coloured)
3. their method of manufacture (pressed, wire cut, handmade)
4. surface texture (rustic, smooth, drag face).

Pressed – regular in shape with sharp arrises with a frog on the top

Wire cut – No frog, sharp arrises, wire cut marks on top and bottom

Handmade – Irregular in shape, shallow frog

Figure 6.7 Methods of brick manufacture

Brickwork NVQ and Technical Certificate Level 2

Find out

What other names of bricks are in common use?

Safety tip

Take care and stand well clear of a crane used for offloading bricks on delivery

Here are some examples of the manufacturers' names that describe their bricks:

- Michelmersh Handmade Reds
- Claydon Red Multi
- Stafford Blue Engineering
- Stratford Sand Faced Brindle.

Storage of bricks

Most bricks delivered to sites are now prepacked and banded using either plastic or metal bands to stop the bricks from separating until ready for use. The edges are also protected by plastic strips to help stop damage during moving, usually by forklift or crane. They are then usually covered in shrink-wrapped plastic to protect them from the elements.

On arrival to site they should be stored on level ground and stacked no more than two packs high, to prevent overreaching or collapse, which could result in injury to workers. They should be stored close to where they are required so further movement is kept to a minimum. On large sites they may be stored further away and moved by telescopic lifting vehicles to the position required for use.

Great care should be taken when using the bricks from the packs as, once the banding is cut, the bricks can collapse causing injury and damage to the bricks, especially on uneven ground. Bricks should be taken from a minimum of three packs and mixed to stop changes in colour, as the position of the bricks during the kiln process can cause slight colour differences: the nearer the centre of the kiln, the lighter the colour; the nearer the edge of the kiln, the darker the colour as the heat is stronger. If the bricks are not mixed, you could get sections of brickwork in slightly different shades; this is called **banding** and in most cases is visible to the most inexperienced eye.

If bricks are unloaded by hand they should be stacked on edge in rows, on firm, level and well-drained ground. The ends of the stacks should be bonded and no higher than 1.8 m. To protect the bricks from rain and frost, all stacks should be covered with a tarpaulin or polythene sheets.

Chapter 6 Handling and storage of materials

Blocks

Blocks are made from concrete, which may be dense or lightweight. Lightweight blocks could be made from a fine aggregate that contains lots of air bubbles. The storage of blocks is the same as for bricks.

Paving slabs

Paving slabs are made from concrete or stone and are available in a variety of sizes, shapes and colours. They are used for pavements and patios, with some slabs given a textured top to improve appearance.

Storage of paving slabs

Paving slabs are normally delivered to sites by lorry and crane offloaded, some in wooden crates covered with shrink-wrapped plastic, or banded and covered on pallets. They should not be stacked more than two packs high for safety reasons and to prevent damage to the slabs due to weight pressure.

Paving slabs unloaded by hand are stored outside and stacked on edge to prevent the lower ones, if stored flat, from being damaged by the weight of the stack. The stack is started by laying about 10 to 12 slabs flat with the others leaning against these. If only a small number of slabs are to be stored, they can be stored flat (since the weight will be less).

Slabs should be stored on firm, level ground with timber bearers below to prevent the edges from getting damaged if placed on a solid surface. To provide protection from rain and frost, it is advisable to keep the slabs undercover, by placing a tarpaulin or polythene sheet over the top.

Paving slabs stacked flat

Paving slabs on pallet

Safety tip

When working with blocks, make sure you always wear appropriate PPE (personal protective equipment), i.e. boots, safety hat, gloves, goggles and face mask

Safety tip

It is good practice to put an intermediate flat stack in long rows to prevent rows from toppling

133

Brickwork NVQ and Technical Certificate Level 2

Stacked kerbs

Stacked lintels

Definition

Vitrified – a material that has been converted into a glass-like substance via exposure to high temperatures

Kerbs

Kerbs are concrete units laid at the edges of roads and footpaths to give straight lines or curves and retain the finished surfaces. The size of a common kerb is 100 mm wide, 300 mm high and 600 mm long. Path edgings are 50 mm wide, 150 mm high and 600 mm long.

Kerbs should be stacked on timber bearers or with overhanging ends, which provides a space for hands or lifting slings if machine lifting is to be used. When they are stacked on top of each other, the stack must not be more than three kerbs high. To protect the kerbs from rain and frost it is advisable to cover them with a tarpaulin or sheet.

Pre-cast concrete lintels

Lintels are components placed above openings in brick and block walls to bridge the opening and support the brick or block work above. Lintels made from concrete have a steel reinforcement placed near the bottom for strength, which is why pre-cast concrete lintels will have a 'T' or 'Top' etched into the top surface. Pre-cast concrete lintels come in a variety of sizes to suit the opening size. The stacking and storage methods are the same as for kerbs.

Drainage pipes

Drainage pipes are made from **vitrified** clay or plastic. Pipes made from clay may have a socket and spigot end or be plain and joined by plastic couplings. Plastic pipes are plain ended and joined by couplings, using a lubricant to help jointing.

Chapter 6 Handling and storage of materials

Storage of drainage pipes and fittings

Pipes should be stored on a firm, level base, and prevented from rolling by placing wedges or stakes on either side of the stack. Do not stack pipes any higher than 1.5 m, and taper the stack towards the top.

Clay pipes with socket and spigot ends should be stored by alternating the ends on each row. They should also be stacked on shaped timber cross-bearers to prevent them from rolling.

Stacked drainage pipes

Remember

Clay pipes are easily broken if misused, so care must be taken when handling these items

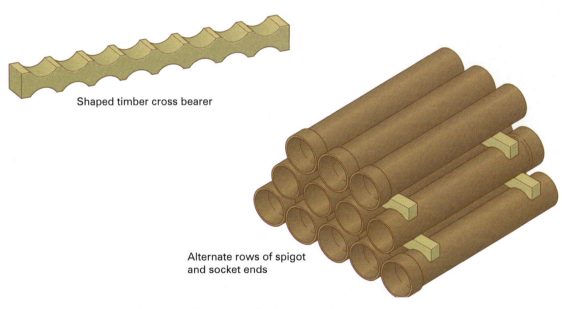

Shaped timber cross bearer

Alternate rows of spigot and socket ends

Figure 6.8 Timber cross-bearer and pipes stacked on cross-bearer

Fittings and special shaped pipes, like bends, should be stored separately and, if possible, in a wooden crate until needed.

135

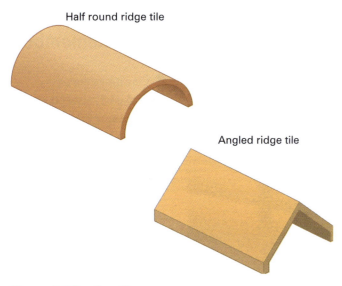

Figure 6.9 Roofing tiles

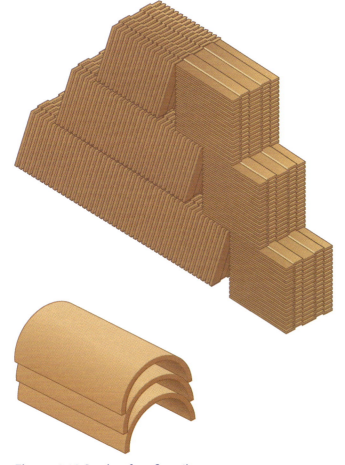

Figure 6.10 Stacks of roofing tiles

Roofing tiles

Roofing tiles are made from either clay or concrete. They may be machine-made or handmade, and are available in a variety of shapes and colours. Many roofing tiles are able to interlock to prevent rain from entering the building. Ridge tiles are usually half round but sometimes they may be angled.

Storage of roofing tiles

Roofing tiles are stacked on edge to protect their 'nibs', and in rows on level, firm, well-drained ground. See Figure 6.9. They should not be stacked any higher than six rows high. The stack should be tapered to prevent them from toppling. The tiles at the end of the rows should be stacked flat to provide support for the rows.

Ridge tiles may be stacked on top of each other, but not any higher than ten tiles.

To protect roofing tiles from rain and frost before use, they should be covered with a tarpaulin or plastic sheeting.

Damp proof course (DPC)

A damp proof course (or DPC) and damp proof membranes are used to prevent damp from penetrating in to a building. Flexible DPC may be made from polythene, bitumen or lead and is supplied in rolls of various widths for different uses.

Slate can also be used as a damp proof course – older houses often have slate but modern houses normally have polythene.

Damp proof membrane is used as a waterproof barrier over larger areas, such as under the concrete on floors etc. It is normally made of 1000-gauge polythene, and comes in large rolls, normally black or blue in colour.

Storage of rolled materials

Rolled materials, for example damp proof course or roofing felt, should be stored in a shed on a level, dry surface. Narrower rolls may be best stored on shelves but in all cases they should be stacked on end to prevent them from rolling and to reduce the possibility of them being damaged by compression. See Figure 6.11. In the case of bitumen, the layers can melt together under pressure.

Plywood and plasterboard

Plywood is made from thin layers of wood glued together with their grains alternating in direction, and is used for flooring, shuttering, and stud partition walling.

Plasterboard is made from a gypsum plaster centre sandwiched between two sheets of heavy-duty paper. It is used for ceilings and stud partition walling.

Both plywood and plasterboard are classed as sheet materials.

Damp proof course (DPC)

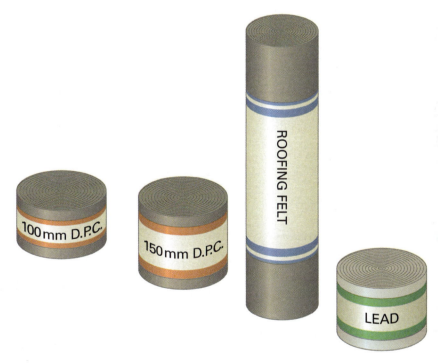

Figure 6.11 Rolled materials stored on end

Flat storage of plywood and plasterboard

Storage of sheet materials

Plywood and plasterboard is best stored flat on racks, in a dry, warm place, but where space is limited sheet materials can be stored on edge in a specially made rack which allows the outer board to be supported on the previous board, and keep its shape. In most cases on sites they are stacked on timber bearers, either flat or on edge.

Liquid materials

In your role as a bricklayer, you will come across different types of liquid material, two of the main ones being plasticiser and brick cleaner.

Plasticiser

Plasticiser is used as an additive for mortar to aerate the mix to improve workability. It comes in liquid form as well as powder. If you have used mortar without plasticiser you will know that it does not hold together when you roll it but dries out quickly and generally seems 'dead'. With plasticiser added it is smoother and far easier to use.

> **Safety tip**
>
> Care must be taken when using plasticiser as it can irritate the skin. You should always read the label for the correct amount to use, as well as all safety instructions and storage requirements

Chapter 6 Handling and storage of materials

Brick cleaner

This is a type of acid used to clean mortar splashes, stains on finished brickwork, and also to clean patios and pathways. It comes in plastic containers and is normally applied by brushing it on to the bricks and then washing off with water. Great care must be taken when using the cleaner not to splash it on to the skin, as it will burn the skin very quickly. Any clothing that comes into contact with it will stain and dissolve, as the material will rot over a short period.

If the skin or clothing is splashed, the area affected should be washed with water straight away. Suitable overalls, gloves and goggles should be worn for protection.

Brick cleaner has a pungent, ammonia smell that can affect breathing or cause headaches and nausea. Therefore a respiratory mask should always be worn.

Care should be taken when storing containers of brick cleaner after delivery so that no unauthorised person comes into contact with them. Containers should be checked periodically to make sure there are no leaks.

Safety tip

As with all chemicals, great care should be taken to ensure different materials do not mix together and containers should always be kept under lock and key

FAQ

What is the most common injury caused by lifting and handling materials?

Back and neck injuries due to lifting heavy loads and lifting any load using a poor manual handling technique. Manual lifting should be avoided where possible and you should always use a safe technique when you do have to lift something. Look after your back.

Knowledge check

1. What is plasterboard made from?
2. What is the best way to carry cement manually?
3. What type of sand should be used in mortar?
4. What is hardcore used for?
5. If you use incorrect lifting techniques, what can happen to you?
6. What are engineering bricks used for?
7. How can you protect materials stored outside?
8. What must be signed when receiving materials on site?
9. Name two types of cement.
10. What are fine aggregates?
11. What is the maximum height allowed for brick packs?
12. What is coarse aggregate used for?
13. State two materials bricks are made from.
14. Name four items of PPE that might be worn when moving materials.
15. What is the chemical reaction that occurs in cement when water is added?
16. Where can sand be obtained from?
17. Name three methods of manufacturing bricks.
18. What is the maximum height at which to store kerbs?
19. What should you do if materials are short on a delivery?
20. Name two types of material that can irritate the skin.

chapter 7

Setting out

OVERVIEW

Setting out refers to the marking out and positioning of a building. It is a very important operation as the setting out of a building must be as accurate as possible. Mistakes made at the setting out stage can prove very costly later. To appreciate the need for careful and accurate setting out, we have to understand and visualise the finished building and its requirements. When setting out a building, you must make sure that it is in the right place, it is level and, in the case of a building, it is square. This chapter will explain the basic rules and methods used when setting out.

The following topics will be covered:

- Health and safety
- Procedures for setting out
- The materials you will need to set out a building
- Step-by-step guide to setting out a building.

141

Health and safety

There are many things to take into consideration during the stages of setting out, from the tools and equipment required, to the actual foundation digging, whether being dug by hand or using heavy plant equipment.

The area where the structure is to be built could have been **barren** for a long time, and may have become a tipping ground for all kinds of rubbish such as glass, metal objects, even hypodermic needles. Therefore, great care should be taken when clearing the area ready to commence setting out.

When putting pegs into the ground, remember that objects could be buried out of sight such as underground pipes and cables.

Methane gas can also be a problem if the ground has been used previously for material excavation, like sand and gravel, and then backfilled over the course of time for house construction. Some of these considerations would probably have been checked for at an earlier stage, but you should always be aware and take precautions such as wearing appropriate personal protective equipment (PPE).

Definition

Barren – land not looked after or used, maybe wasteland

Did you know?

Previous checks would be carried out by ground survey before site commencement

On the job: Clearing the site

Steven is clearing a newly acquired site when he comes across a pipe sticking out of the ground. There is a sort of open top to it. What do you think this could be? Depending on your answer, what do you think Steven should do next.

Chapter 7 Setting out

Procedures for setting out

Setting out a building is an extremely important operation as mistakes at this stage would be very costly to put right later. Great care should be taken on reading the drawings and making sure that the correct measurements are used.

Setting out at the right place

Finding the right place to set out sounds like an obvious thing to state, but it is so important that this is done absolutely correctly. Buildings have sometimes had to be completely demolished for being put up in the wrong place! This is because **building lines** and **boundary lines** are often involved and there are very strict regulations governing these.

Remember

Incorrect measurements could mean digging out finished concrete

Definition

Building line – an imaginary line set by the local authority to control the positioning of buildings

Boundary lines – lines that dictate ownership, e.g. lines between properties

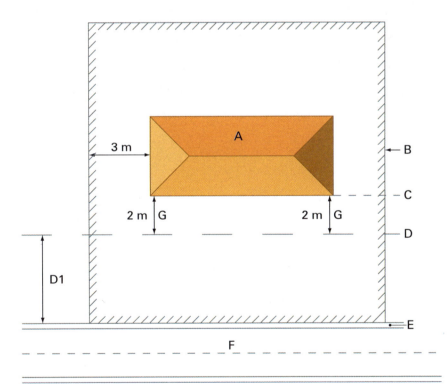

Key
A = Plan of proposed building
B = Boundary of site
C = **Frontage line**
D = Building line
D1 = Distance set by local authority (council)
E = Pavement
F = Road
G = Distance from building line to frontage line

Figure 7.1 Positioning a building in the right place (site plan not to scale)

Definition

Frontage line – the line of the front of the building

Note

The building shown in Figure 7.1 is 3 m from the left hand boundary and 2 m behind the building line – it is most important that a building is accurately positioned

143

Keeping level

Generally speaking most buildings are built level (horizontal) so it makes sense to do the setting out on a level plane. A building is kept level by various methods. Larger buildings are very often positioned in relation to the Ordnance Survey Bench Mark (OSBM) system. Figure 7.2 shows the position of the average sea level of the UK (which is the average level of the sea in Newlyn, Cornwall) and this is called 0.000 metres. Bench Marks are

Figure 7.2 Ordnance Datum Newlyn (ODN)

Figure 7.3 Example of a Bench Mark

positioned countrywide (shown dotted) in relation to this sea level and are often carved in stone walls, churches, government buildings, or sometimes cast in a concrete pedestal as shown in Figure 7.3.

When the Ordnance Survey System is used for reference, the nearest OSBM is used and a relative **Temporary Bench Mark (TBM)**, or site datum, is positioned on site as shown in Figure 7.4. It can be in the form of a steel or wooden peg set in concrete and very often protected by a fence.

Definition

Temporary Bench Mark (TBM) – a fixed levelling point to which other levelling points are related. It is used on larger sites

Chapter 7 Setting out

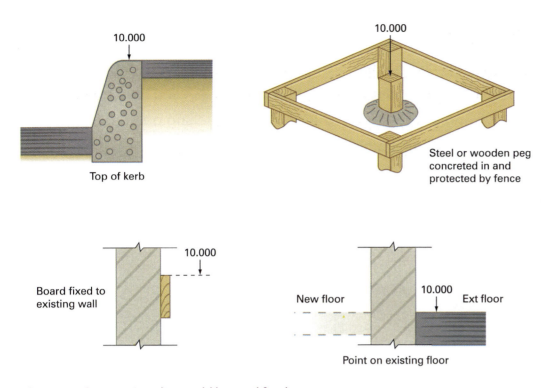

Figure 7.4 Some points that could be used for datums

> **Did you know?**
>
> On smaller jobs the OSBM system is not used so an assumed height is used, say 10.000 m as shown in Figure 7.4. Heights above and below this level are calculated accordingly

Figure 7.5 shows a datum with an 'assumed' level of 10.000 m. The floor level is 10.000 m so the floor level is the same as the datum. The bottom of the excavation is 9.000 m, so the excavation is 1.000 m below the datum peg. The top of the foundation concrete is 9.150 m, so the thickness of the foundation concrete is 9.150 – 9.000 = 0.150 m.

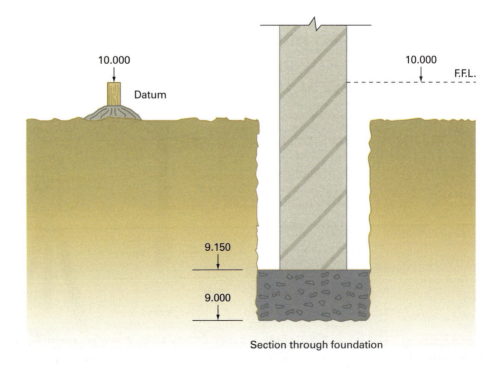

Figure 7.5 Datum with levels

145

> **Remember**
>
> A level can go out of true or a straight edge can be bent – check first

Transferring levels by spirit level and straight edge

This is a very basic method of transferring a level (see Figure 7.6). Peg B is levelled from peg A (the datum) and then peg C is levelled from peg B. The straight edge and level are rotated 180° at each levelling point to eliminate any error in the level or straight edge.

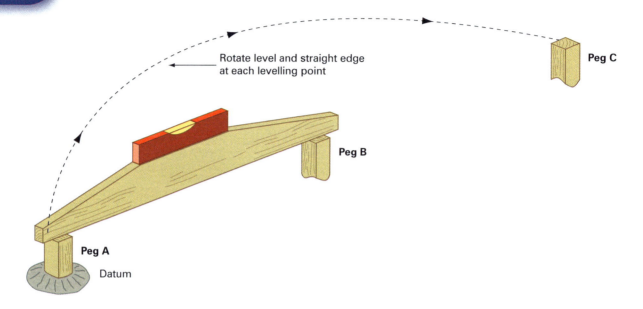

Figure 7.6 Transferring levels

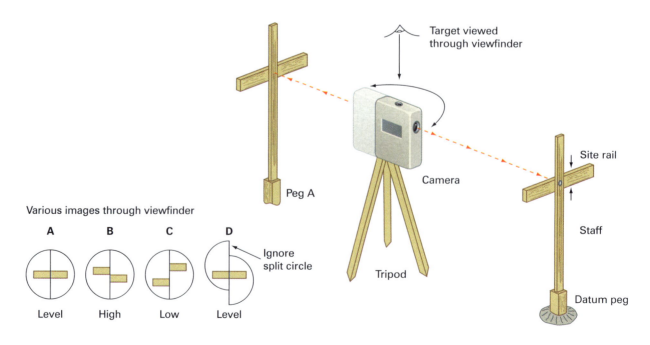

Figure 7.7 Cowley optical level

Optical level

An optical level is a levelling device that comprises a camera, a tripod and a staff (target). The camera swivels 360° on a pin and is accurate up to about 30 metres. This type of level is called a Cowley or Quickset level (see Figure 7.7).

Procedure for levelling with an optical level

1. Securely set up tripod and lower level on to pin.
2. Position staff on datum and observe through viewfinder.
3. Move site rail up or down to obtain image A or D.
4. Secure site rail with screw.
5. Place staff on peg A and rotate level to observe.
6. Adjust height of peg to obtain image A or D.
7. Peg A will then be level with datum peg.

Remember

The tripod should not be disturbed during levelling. To avoid internal damage to the camera, never carry the level while on the tripod

Dumpy level

A dumpy level is an optical level set on a tripod, the same as the Cowley, but is more accurate and sets levels over a greater distance. This type of optical level can give an accurate reading up to approximately 150 metres away to a tolerance of 3 mm. On looking through the telescopic site the fixed target site is read against the staff on the datum, and all other levels are adjusted to suit.

How to set up a dumpy level

1. The legs on the tripod are extended to suit the individual user so that the tripod plate is about chin height. This stops any reaching up or bending which could result in the tripod being knocked, with a false reading being taken.
2. Fix the level to the tripod using the centre screw.
3. Adjust the three individual screws at the base to set the top bubble into the centre of the levelling circle.

Dumpy level

Laser level

4. Rotate the level 360° to check that the bubble stays in the circle; if it does not, then fine adjustment is required until correct.

5. Once the telescopic site is level, locate the staff set on top of the datum. You may have to adjust the focus using the large adjuster on the side of the level. Once you have located the staff, final adjustment may be required to see the stadia lines of the target cross. This is achieved by adjusting at the actual eye piece, turning to the left or right.

6. The level is now ready to use, with another person holding the staff as upright as possible for an accurate reading to be taken from the staff.

Laser level

Laser levels are the newer modern technology in construction, taking over from the Cowley and dumpy levels. They are very accurate and easy to set up. The level is fixed to a tripod. Press the button and it automatically finds level, shooting a red dot that can be picked up on the staff giving the reading. Laser levels can be used for all types of levelling throughout the course of the work, from setting foundations and floor heights, to suspended ceiling levels or even putting in straight plumbing pipework. Some are accurate up to 100 metres and, if the level is knocked, it gives notification of movement.

Keeping square

It is most important that a building has square (90°) corners. Therefore, the setting out of a building must be square to avoid construction problems later, e.g. the roof not fitting! There are sometimes exceptions, as in a circular building or a building on an awkwardly shaped site.

Site square

A site square is an optical instrument used to set out right angles simply and accurately. It is supported by a tripod, and the body of the instrument contains two telescopes which are mounted at 90° to each other. The telescopes also pivot vertically so various distances can be sited.

Site square

A lock screw is located under the body of the instrument, and a fine adjustment screw enables accurate location of the target. The site square is plumbed by a bubble in a circle levelling device on top of the instrument.

Procedure for using a site square

1. Set up datum rod A over corner nail.
2. Site nail in Peg B using lower telescope.
3. Position nail in Peg C using upper telescope.

Builder's square

A builder's square is the most commonly used way of setting out a corner if no optical square is available on the site. It is made of timber, is usually made on site and obviously has to be practical enough to be easily carried and held in place for checking corners.

Remember

Manufacturers' instructions will explain in more detail how to use equipment and check for accuracy

How to set out a right angle using a builder's square

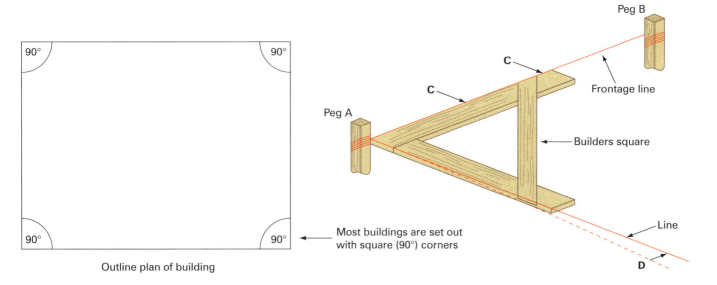

Figure 7.8 Right angle set out with a builder's square

1. Set out Frontage Line Peg A to Peg B.
2. Position square parallel with Frontage Line C.
3. Adjust line at D to make a right angle at the corner.
4. When the line is to the side of the square, the corner is square (90°).

The 3:4:5 method

A building can also be set out using the 3:4:5 method, which is a way of forming right angles using **trigonometry**. It may sound it, but the 3:4:5 rule is not very difficult to understand. If you take three straight lines, one 3 cm long, one 4 cm long and one 5 cm long, and then join them together to make a triangle, the angle opposite the longest line will always be a perfect right angle.

Definition

Trigonometry – the part of mathematics concerned with triangles and angles

How to set out a right angle using the 3:4:5 method

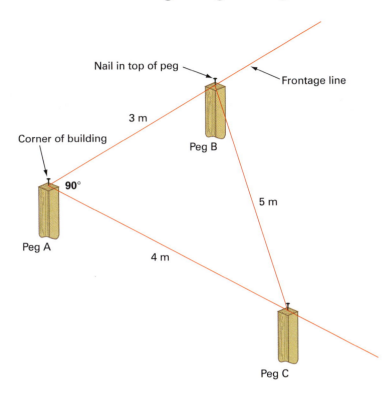

Figure 7.9 Right angle set out using 3:4:5 method

1. Set up Frontage Line from Peg A.
2. Position Peg B.
3. Set up Peg C.
4. Right angle formed as shown.

Did you know?

The ancient Egyptians used the 3:4:5 method to build the pyramids about 4,500 years ago

Remember

Always use the largest possible triangle for accuracy. Triangles can be any size as long as the ratio is 3:4:5

On the job: Checking corners for square

Jack has been asked to recheck a corner to make sure it is a right angle. He has found that the optical square and builder's square have been taken to another site to be used. How can he check the corner without these?

The materials you will need to set out a building

The exact materials you will need to set out a building will vary depending on the size of the job. The following list would be adequate for, say, a small detached house:

- plans and specifications
- two measuring tapes (30 m), preferably steel
- optical level
- site square (optional)
- 50 mm x 50 mm timber pegs
- 25 mm x 100 mm timber for profiles
- lump hammer
- claw hammer
- hand saw
- a line
- concrete (ballast and cement) to secure pegs (although sometimes unnecessary)
- sand (for marking out trenches)
- 50 mm round head nails
- 75 mm round head nails.

The list is not exhaustive but should give you some idea of what is required. After checking on the size of the job, you should adjust this list as appropriate.

Step-by-step guide to setting out a building

We will look at the steps involved in setting out over the next few pages. Listed below are a few golden rules you should always observe during the setting out process:

1. Make sure you know where the building line and boundaries are.
2. Check your equipment before commencing.
3. Establish a datum where it will not be disturbed.
4. Always use measurements given and avoid scaling.
5. Set out a base line (e.g. front of house). Make sure you do not infringe on or over the building line.
6. Be aware of any underground pipes etc.
7. Check the drawings for errors.
8. Take all measurements with care and accuracy.
9. Check and double check setting out after completion.

Step 1 Establish front base line

(i) Peg A to Peg B.

(ii) Nails indicate corners of building.

(iii) Pegs reasonably level with each other.

(iv) Pegs must be secure and not move.

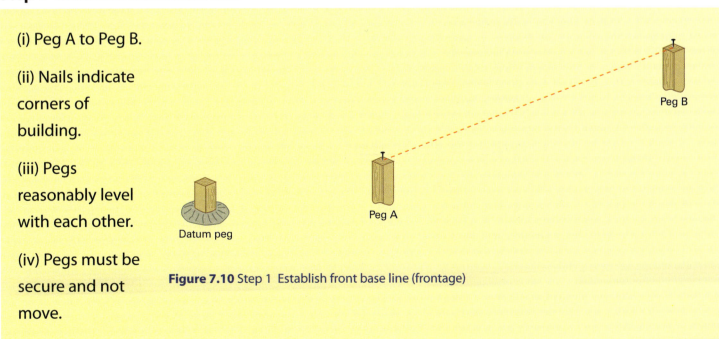

Figure 7.10 Step 1 Establish front base line (frontage)

Brickwork NVQ and Technical Certificate Level 2

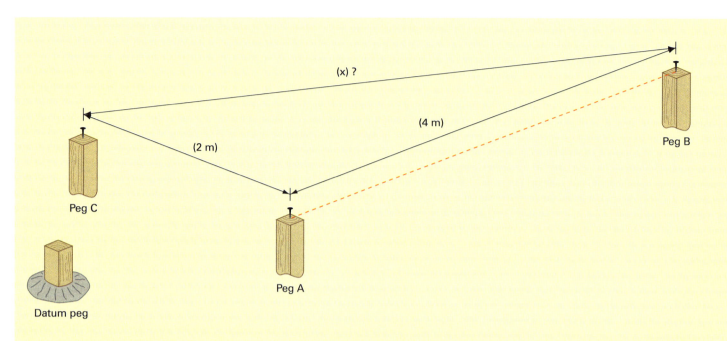

Figure 7.11 Step 2 Establish Peg C (using a site square, builder's square or the 3:4:5 method)

Step 2 Establish Peg C

Assume sizes are

4 m x 2 m:

$X^2 = 4^2 + 2^2$

$X^2 = 16 + 4$

$X^2 = 20$

$X = \sqrt{20}$

$X = 4.472$ m

Two tapes can be used now from A and B to find C.

Step 3
Establish Peg 3

(i) Use two tapes and measure from Pegs C and B.

(ii) Check diagonals A–D and C–B.

(iii) Building is square if diagonals are the same.

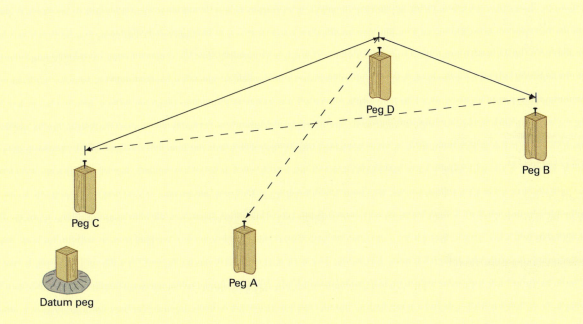

Figure 7.12 Step 3 Establish Peg 3

Step 4 Erect profiles at E and F

(i) Project line from nails in Peg A and B.

(ii) Mark profiles with nail or saw cut.

(iii) Profiles should be 1 m minimum away from face – further if machine digging.

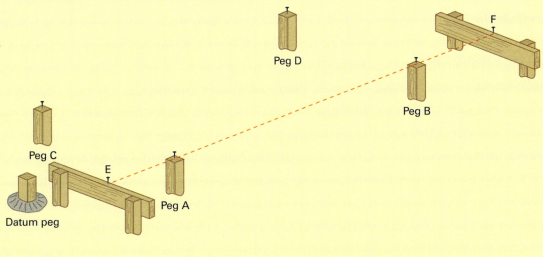

Figure 7.13 Step 4 Erect profiles at E and F

Step 5 The **profile** at Peg D showing alternative.

> **Definition**
>
> **Profile** – a support for a line outside the working area

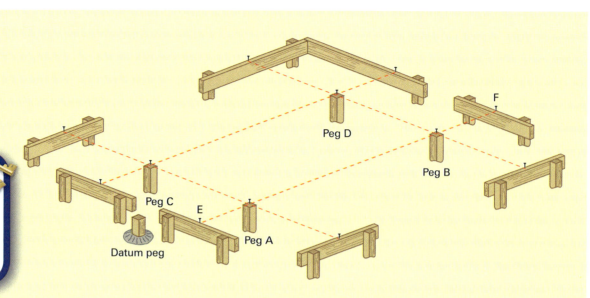

Figure 7.14 Step 5 Repeat step 3 for remaining profiles

Step 6 Remove corner pegs

(i) Attach continual line as shown.

(ii) Line represents face line(s).

(iii) Line should not 'bind' on crossing.

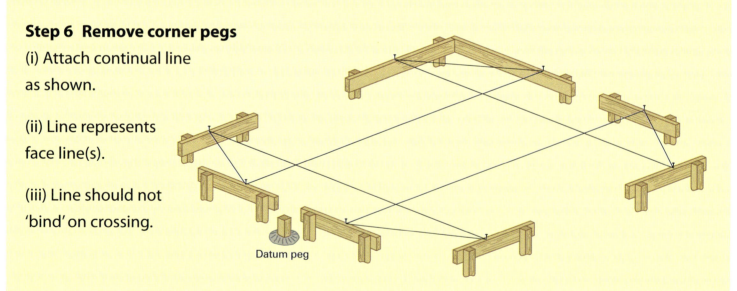

Figure 7.15 Step 6 Remove corner pegs

Step 7 Edges of foundations marked on profiles

(i) Line attached – plumbed down and marked on ground with sand (shown dotted).

(ii) Trenches excavated.

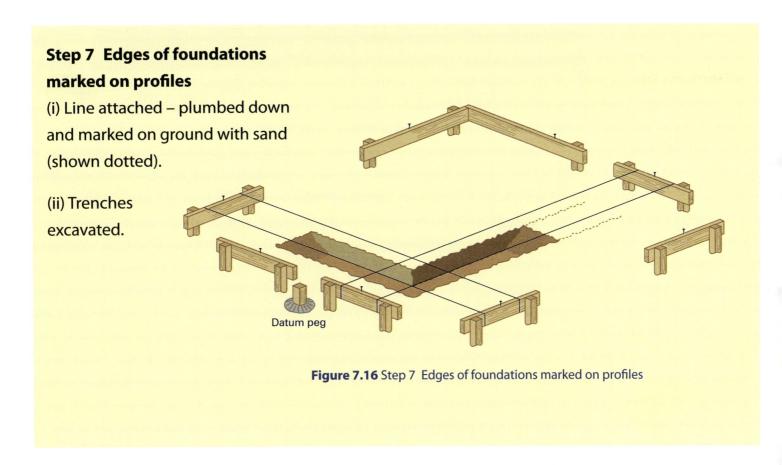

Figure 7.16 Step 7 Edges of foundations marked on profiles

FAQ

With so many different types of level available, which is the best type to use when setting out?

A laser level is probably the best type of level currently available (see page 148 for why), however, you may be limited by what equipment is actually on site. A dumpy level is probably the next best piece of levelling equipment available. Whatever you use, make sure you check it first and use it properly.

Knowledge check

1. What type of optical level can you set out a corner with?
2. What is another name for a temporary bench mark?
3. Where is the average sea level taken from when setting out?
4. Why do you reverse the level and straight edge when levelling?
5. Why should you be careful when excavating ground previously used for gravel works?
6. What are the two standard optical levels most commonly used?
7. What does OSBM stand for?
8. Why should you ensure a building is square when setting out?
9. Other than an optical square, what else can be used to set out a corner?
10. What are the golden rules you should observe when setting out? Don't forget what order they are in!

chapter 8

Mixing mortar

OVERVIEW

Mortar is used in bricklaying for bedding and jointing bricks and blocks when building walls. This chapter will explain how to mix mortar to different strengths for a number of different tasks.

This chapter will cover:

- Mortar
- Choice of mortar mixes
- Gauging materials
- Mixing of materials
- Pre-mixed mortars.

Mortar

Mortar is used in bricklaying for bedding and jointing the bricks when building a wall. Mortar is made of sand, cement, water and plasticiser. The mortar must be 'workable' to allow the mortar to roll and spread easily. The mortar should hold on to the trowel without sticking.

Sand

Sand for bricklaying mortar should be 'well graded', having large, medium and small grains (see Figure 8.1). If all the grains were of even size this would be termed 'poorly graded' and would require more cement to fill in the voids between each grain. Sand is dug from pits or dredged from the sea and in both cases must be thoroughly washed to remove mud and silt. A way to check for the cleanliness of sand is to carry out a silt test.

> **Did you know?**
>
> On cold mornings, brickwork should not be started unless the temperature is 3°C and rising

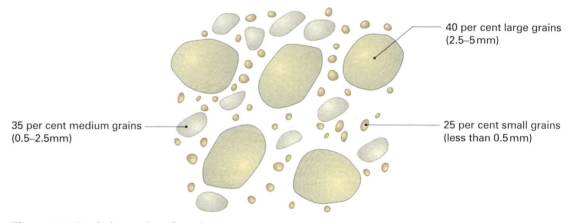

Figure 8.1 Graded samples of sand

Silt test

The silt test is an easy method to determine the amount of mud and silt that is mixed in with the sand once it arrives on site. It will also show the amount of clay left in the sand, as this will stop the mortar bonding to the sand particles. The amount of silt and clay should not be more than 10% of the volume of aggregate.

See Chapter 16 Basic concreting (page 279) for how to carry out a silt test.

Cement

Cement is made from limestone or chalk and chemically controlled with added calcium, aluminium, silicon and iron. It takes 1.65 tonnes of raw material to make 1 tonne of cement, with half of the weight of the limestone material being lost through carbon dioxide emissions during the manufacturing process. The materials are then proportioned to form a raw mix and introduced into a kiln.

There are four stages within the kiln:

1. evaporation and preheating

2. calcining

3. clinkering, and

4. cooling.

The first stage removes moisture and then raises the temperature of the material ready for calcining to between 800 and 900°C, which breaks down the calcium carbonate into calcium dioxide and carbon dioxide. This process produces a substance known as clinker. This is then ground into a powder and mixed with gypsum to produce the finished cement powder. The gypsum controls the rate of hydration of the cement in the setting process.

Cement is used to bind the grains of sand together. A layer of cement slurry coats the particles of sand, which chemically sets after the addition of water, resulting in a hardened layer holding the bricks in place. The most common cement used is Ordinary Portland Cement (OPC), which is suitable for most general work and, if handled correctly, will produce mortar of a high quality and strength.

Masonry cement is often used for bricklaying mortar. It is similar to OPC but has a plasticiser added to the cement powder. As a bag of masonry cement contains 75% cement powder and 25% plasticiser, a higher proportion of cement must be used.

Sulphate-resisting cement is suitable for use below ground, where high levels of sulphate may damage Ordinary Portland Cement. It is normally used for foundation works and drainage.

For other types of cement, see Chapter 16 Basic concreting, page 279.

Cement is tested by British Standards to ensure that it is suitable for use. It is then given a Kite Mark and a number to show that it has passed the test. Part of the test specifies the setting time for cement as being not less than 45 minutes before its initial set has started, and not more than 10 hours before its final set has taken place. The cement must be used before its initial set has taken place as any remixing or movement will result in the cement mortar not bonding or setting properly.

On the job: Mixing mortar above ground level

Seb is mixing mortar for six bricklayers building the cavity brickwork on the second floor of a block of flats. They are using Ordinary Portland Cement. He suddenly realises that he is down to the last bag. There are eight bags of sulphate-resisting cement in the shed. Should he use these?

Water

Water is used to make the cement paste and to cause the cement to set due to a chemical reaction (**hydration**). The water used for mixing must be clean enough to drink.

Definition

Hydration – the addition of water to cement paste to produce a chemical reaction to set mortar

FAQ

I've heard that mortar can be made up without cement and just have lime added to it – why is this?

Before Portland cement (OPC) was introduced, mortar was made from sand, slaked lime and water. Mortar made without OPC and just the addition of lime is often used in restoration work. This is because many old buildings were not built with OPC and so mortar made with OPC will be stronger than the bricks it binds.

Plasticisers

Sand, cement and water will make a mortar that will be difficult for a bricklayer to use. To make a bricklaying mortar 'workable' a plasticiser must be added. Most plasticisers nowadays come in a powder or liquid form, which should be added to the water according to the instructions. The plasticiser works by coating the grains of sand with tiny bubbles of air, which allows the sand to flow easily when being spread.

Hydrated lime may be used as a plasticiser but washing up liquid etc. must *never* be used as the amount of air bubbles cannot be controlled and would result in a weak mortar mix as the chemicals and detergents react to the cement, breaking down the hydration.

Remember

Never use plasticiser below ground level

Colouring agents

Colouring agents are available in powder or liquid form but are only really suitable for pointing brickwork as it is almost impossible to keep a consistent colour in a large amount of mortar by this method. If powder pigments are to be used for pointing, it is recommended to mix enough mortar to point the whole of the work dry, and keep stored in airtight bags in a dry store.

Where large amounts of coloured mortar are required it is recommended to use a ready mix mortar, which is available from mortar suppliers.

Remember

When using colouring pigments, always follow the instructions on the packaging

Choice of mortar mixes

The choice of the mortar to be used for a project depends on:

- the cost
- exposure
- weather conditions
- type of brick to be used.

A general rule is that the mortar must not be stronger than the bricks to be used. This enables a wall to be simply repointed if there are any cracks in the joints due to settlement. If the mortar were stronger than the brick, the settlement crack would go vertically through the brick, resulting in the brickwork having to be rebuilt.

Mortars are usually described as a ratio of materials, for example 1:1:5. The first number is always the proportion of cement, the second is the proportion of lime and the third number is the proportion of sand.

Many mortar mixes are given a 'designation' number (i, ii, iii, iv, v). This allows different batches of mortar to be approximately the same strength. The lower the designation number (i) the stronger and more durable the mortar, while the higher the number (v) the greater the ability of the mortar to allow for movement of the wall.

Designation	Type		
	OPC:Lime:Sand	Masonry cement: Sand	OPC: Sand with plasticicser
v	1:3:10–1:3:12	1:6½–1:7	1:8
iv	1:2:8–1:2:9	1:5½–1:6½	1:7–1:8
iii	1:1:5–1:1:6	1:4–1:5	1:5–1:6
ii	1:½:4–1:½:4½	1:2–1:3½	1:3–1:4
i	1:0:3–1:¼:3		

Table 8.1 Mortar mix designations

Gauging materials

Accurate measuring of materials to the required proportion before mixing is important to ensure consistent colour, strength and durability of the mortar. The most accurate method of gauging the mortar materials is by weight, although this method is usually only used on very large sites. The next best way to gauge the materials is by volume using a **gauge box**.

A gauge box is a bottomless box made to the volume of sand required (to a proportion of a bag of cement). The box is placed on a clean, level, flat surface and filled with the sand. The sand is levelled off and any spillages cleaned away. The box is then removed, leaving the amount of sand to be mixed with the bag of cement. If a gauge box is not available, a bucket could be used. A bucket is filled with sand and emptied on a clean, flat surface for the number of times specified in the proportion. A separate bucket should be used to measure the cement.

Mixing of materials

Mixing by hand

If mixing by hand, the materials should be gauged first into a pile with the cement added. The cement and sand should then be 'turned' to mix the materials together. The pile should be turned a minimum of three times to ensure the materials are mixed properly. The centre of the pile should be 'opened out' to create a centre hole.

Gradually add the water, mixing it into the sand and cement, making sure not to 'flood' the mix. Turn the mix another three times, adding water gradually to gain the required consistency.

> **Remember**
>
> Make sure buckets used to gauge materials are the same size

> **Did you know?**
>
> A 'shovel full' of material is not a very accurate method of gauging materials and should be avoided

> **Remember**
>
> It is easier to add more water than to try to remove excess water

Chapter 8 Mixing mortar

165

Mixing by machine

Mixing by machine can be carried out by using either an electric mixer or a petrol or diesel mixer. Always set the mixer up on level ground.

If using an electric mixer the voltage should be 110 V and all cables and connections should be checked before use for splits or a loose connection. Cables should not be in contact with water and the operation should not be carried out if it is raining.

If using a petrol or diesel mixer, make sure the fuel and oil levels are checked and topped up before starting. If using the mixer for long periods, the levels should be checked regularly so as not to run out.

Gauge the materials to be used, fill the mixer with approximately half of the water required (add plasticiser if being used). Add half the amount of cement to the water and add half of the sand. Allow to mix, and then add the remaining cement, then sand. Add more water if required, allowing at least two minutes for the mix to become workable and to ensure all the materials are thoroughly mixed together.

Once the mix has been taken out of the mixer, part fill the mixer with water and allow the water to run for a couple of minutes to remove any mortar sticking to the sides. If the mixer will not be used again that day, it should be cleaned thoroughly, either using water (and adding some broken bricks to help remove any mortar stuck to the sides) or ballast and gravel (which should then be cleaned out and the mixer washed with clean water). This will keep the mixer drum clean, and any future materials used will not stick to the drum sides so easily.

Remember

Never hit the drum with a hammer etc. to clean it out as this could result in costly repairs to the drum

Pre-mixed mortars

On most larger sites, mixing mortar is a thing of the past as it is brought in already mixed. There are many advantages to using pre-mixed mortar:

- The mix is always consistent and not affected by the weather.
- The mix has better productivity.
- The mix is site efficient.
- There is less site activity transporting materials.
- There is less waste and contamination.

There are two main ways that pre-mixed mortars are used:

1. Mortar silo.
2. Ready spread tubs.

Mortar silo

A mortar silo is a holding unit for mortar. The silo is delivered to the site and has the advantage of taking up little space and preventing contamination of materials (which means waste).

Double compartment silo

The silo is delivered to site and has a mixing unit inside. It is then connected to a power and water supply. The silo is also split into two separate compartments: one is normally filled with a sand-lime mix to the required mix as specified, and the other with cement. The silo is usually filled before transport to site. The mix ratio is calibrated before delivery so that the mixture is exactly the same for each mix.

Liquid admixtures or colours can be added according to the customer's requirements before transportation.

Single compartment silo

This type of silo only has a single compartment which is filled by the supplier with dried sand, cement and lime (if required), plus admixtures or colour pigment as required, with water added for the required consistency.

Ready spread tubs

These are ready mixed plastic tubs containing the required mix of mortar. They have a life span of approximately two days after delivery. The plastic tubs are returnable and carry enough mortar to lay about 350 bricks. They normally have a plastic cover to stop contamination and the air drying them out too quickly. These are very versatile as they can be lifted by forklift straight on to a scaffold for use. They also come in a wide range of colours. There is no site mixing and labour costs are reduced.

Knowledge check

1. What is the difference between masonry cement and Ordinary Portland Cement?
2. What standard should water be for mixing mortar?
3. What two main materials is cement made from?
4. What are the main reasons for gauging materials?
5. When mixing mortar, what is the purpose of plasticiser?
6. What type of sand should be used for mortar?
7. What is hydration?
8. Where would you use sulphate-resisting cement?
9. What is the most commonly used type of cement?
10. What test is carried out to check the amount of mud etc. in sand?

chapter 9

Bonding

OVERVIEW

Bonding is the term given to the different patterns produced when building brick walls. The main reasons for bonding brickwork are for strength, to distribute heavy loads and to help resist sideways and downward pressure to the wall.

The majority of brickwork is built in ½ brick walling in the form of cavity wall construction, using stretcher bond as this is the most economic bond. Due to the wide and varied work carried out by a bricklayer, other types of bond are used. This chapter examines the main bonds used in everyday work, but there are many more.

This chapter will cover:

- What is bonding?
- Types of bond
- Block work.

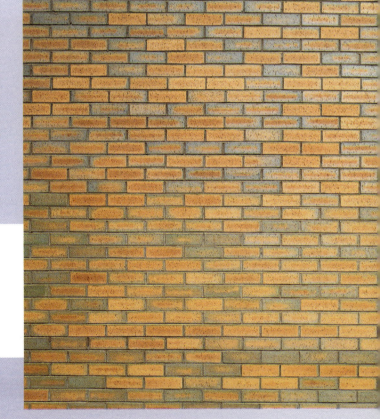

What is bonding?

Did you know?

Bricks laid without bonding are called straight joints

Bonding is the lapping of bricks to gain the most strength from the finished item for a particular job. The lapping is carried out in two ways: half brick lap and quarter brick lap but these are better known as half bond and quarter bond. If bricks were just put one on top of the other in a column, there would be no strength to the wall, and with sideways and downward pressure this type of wall would just collapse.

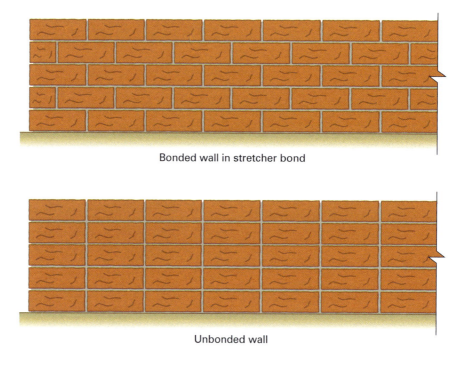

Figure 9.1 Bonded and unbonded walls

Before carrying out any bonding, you must understand the basic sizes and names of the different parts of a brick, and the cuts used to enable you to set out a wall correctly.

Chapter 9 Bonding

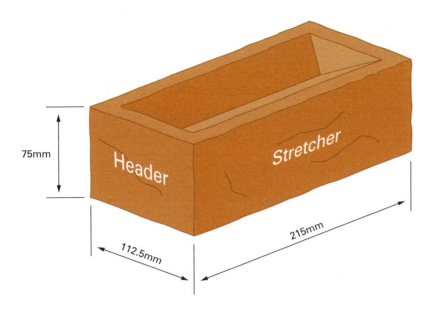

Figure 9.2 Sizes of brick, header and stretcher

Remember

There can be quite a big difference in the sizes of bricks

As you can see in Figure 9.2, the length of a brick is called a **stretcher** with the end of a brick called a **header**. The length of the stretcher is 215 mm and the header length is 102.5 mm. Two headers plus a 10 mm joint equal the length of the stretcher. These measurements can vary slightly as all bricks are not exactly the same and depend on the mould sizes used when produced at the factory, meaning the joint size may have to be adjusted.

Now you understand the basic sizes, you can start to set out your wall. Remember that in most cases the wall to be built is governed by measurements given on a drawing. These do not always work to brickwork sizes, therefore always set out the wall dry, bonding in the required bond using a stretcher course to establish if cut bricks will need to be used. If a cut is required it should be placed in the centre of the wall, with the smallest cut being a half brick (102.5 mm). Sometimes the cut can be put under a door or window, or a reverse bond may be another option (refer to Chapter 11 Solid walls for more information).

Note

If a quarter of a brick is to be gained, use ½ brick and ¾ brick to fill

171

Types of bond

There are many different types of bond that can be used. The choice of bond is usually determined by the purpose of the wall and the strength required, incorporating the thickness.

Half bond

As mentioned previously, half bond is the most common bond and is used mainly in the construction of the outer leaf in cavity wall brickwork (see Figure 9.3) and the inner leaf in block walling (for more information see Chapter 12 Cavity walling, page 207).

Sometimes you may come across a project where the outer leaf of brickwork may be built using a bond other than stretcher bond. In most cases this is when an extension is carried out to an older property and the brickwork needs to blend with the existing bond.

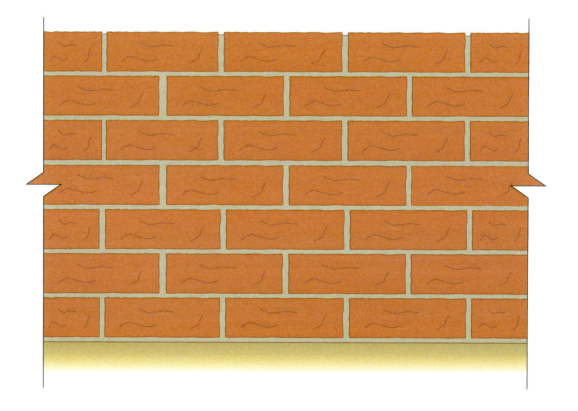

Figure 9.3 Half bond brickwork (cavity wall)

Quarter bond

Quarter bond is used in the construction of solid walling used for garden walls, load bearing walls and retaining walls, as well as manholes, that are 215 mm thick or above. It is used for its strength and, on some work, for its appearance. There are many different types of quarter bond but there are four main bonds that are regularly used:

1. English bond
2. English garden wall bond
3. Flemish bond
4. Flemish garden wall bond.

English bond

English bond is the strongest of all the bonds using alternate courses of stretchers and headers (see Figure 9.4). Used for manholes, garden walls etc. its appearance can look monotonous but strength is the priority.

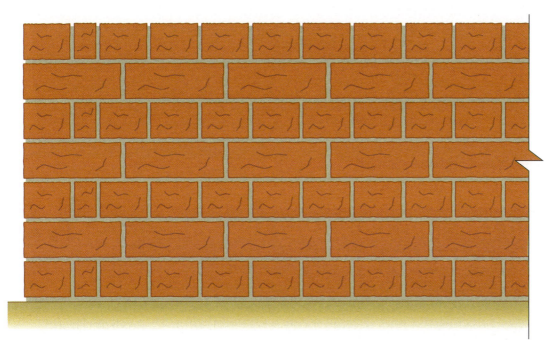

Figure 9.4 English bond

English garden wall bond

English garden wall bond consists of three courses of stretchers and then one course of headers (see Figure 9.5). This bond is not as strong as English bond as there is a straight joint in the centre of the wall on the three stretcher courses. As the name suggests, English garden wall bond is mainly used for garden wall construction as downward pressure is not a problem.

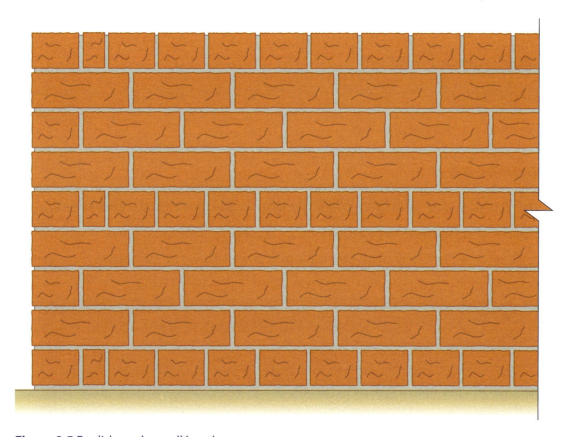

Figure 9.5 English garden wall bond

On the job: Choosing the bond for a garden wall

Dewi needs to build a wall in his garden to retain the earth to a raised area. The difference in the levels is 680 mm. What size wall and bond would be best suited for this job? What could Dewi do to deal with any water that collects from the raised area? What should Dewi do about protecting the top of the wall from weathering?

Flemish bond

Flemish bond uses alternate stretchers and headers in each course (see Figure 9.6). The header should be positioned in the centre of the stretcher of the course below and above, making it the most attractive bond used, especially if the headers are carried out in an alternative coloured brick.

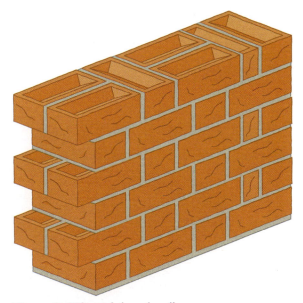

Figure 9.6 Flemish bond wall

Flemish garden wall bond

Flemish garden wall bond consists of three stretchers and then one header alternating in each course (see Figure 9.7). The header is positioned in the centre of the middle stretcher each time.

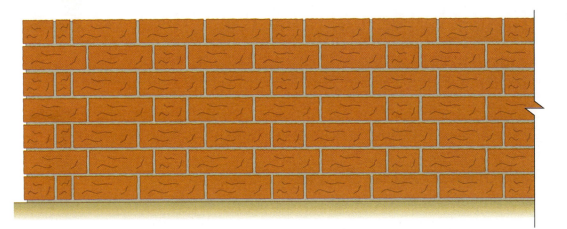

Figure 9.7 Flemish garden wall bond

Did you know?

Flemish garden wall bond is stronger than English garden wall bond as the headers tie across more regularly

Chapter 9 Bonding

175

Brickwork NVQ and Technical Certificate Level 2

Remember

For bonding purposes never use a closer in the wall other than at the corner

In all cases where quarter bond is used, the corner header should have a queen closer next to it to form the quarter bond (see Figure 9.8).

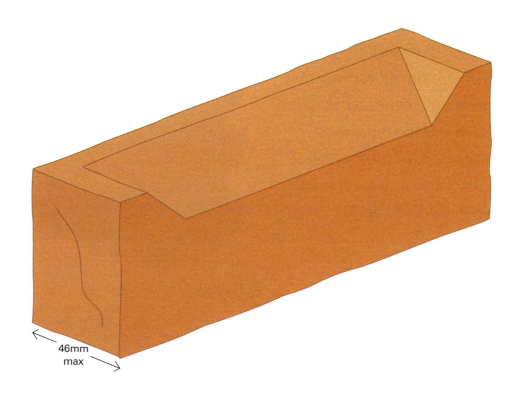

Figure 9.8 Queen closer – a brick that is cut the full length by just under half the width (46mm max) to allow for a joint

On the job: Choosing the bond for a cavity wall

Neil is setting out a cavity wall for a new extension. The existing brickwork is in Flemish bond. He starts setting out in stretcher bond as the bricks add up to the same measurements, i.e. 2 half bricks = 1 stretcher. Is this OK?

Block work

When building block work, the same principles apply as with half bond brickwork. It should be set out to use as many full blocks as possible. At an internal corner, a 100 mm section of block should be used to gain half bond. Never use brick on an internal corner, as the thermal value of a brick is not to the same standard as a block and will cause a 'cold spot' to the finished area possibly showing through the plaster on completion.

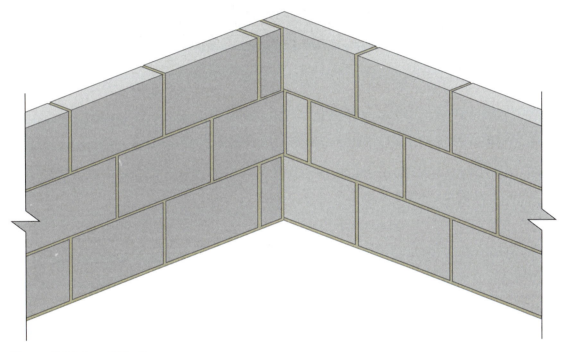

Figure 9.9 Internal block corner

Walls above 215 mm thick

Because of the cost of bricks, most walls above 215 mm thick usually incorporate blocks as they are more cost effective but have the strength required. If used for external works, concrete or similar density blocks should always be used. The walls should either be bonded to each other or connected by means of wall ties. External block work used for strength on work like retaining walls can be laid flat for extra strength, with a brick face for finish.

Did you know?

Walls over two bricks thick are now more likely to be built using solid concrete, formed and shuttered in place to be cost effective (see also Chapter 11 Solid walls, page 191)

FAQ

What is the most commonly used bond for 1 brick walls?

English bond is used most often when constructing a 1 brick wall as this type of bond is the strongest.

Knowledge check

1. What is bonding?
2. What is the length of a brick called?
3. What is the end of a brick called?
4. What bond consists of three stretchers and one header alternating in a single course?
5. Give one of the main reasons for bonding bricks or blocks.
6. What type of wall could be built using English garden wall bond?
7. What is the smallest cut that can be used in a wall?
8. What is the main bond used in cavity work?
9. What should be used next to a header on a corner to gain bond?
10. What bond uses alternative courses of stretchers and headers?

chapter 10

Laying bricks and blocks to a line

OVERVIEW

Bricklayers need to build walls level (flat), plumb (upright) and straight (face plane), as well as finishing the wall to give a pleasing appearance. These skills do not happen overnight – they take time and a lot of practice as well as following the correct procedures and tried and tested methods used for years. This chapter is designed to show the basic procedures of how to lay bricks and blocks to a line.

In this chapter we will cover:

- Health and safety
- Laying bricks
- Laying blocks.

Health and safety

Bricks can vary in weight depending on the type. For instance, engineering and concrete bricks are heavier than clay bricks. Water absorption to the bricks also increases the weight. The same applies for blocks.

You must always remember that bricks and blocks are a potential source of danger when handling them in your daily work. When lifting bricks and blocks, always use correct methods, i.e. **kinetic lifting** (see Chapter 6 Handling and storage of materials, page 122 for more information). Heavy dense concrete blocks can cause back and hand injuries when lifting, and if dropped on your foot or hand can cause serious damage.

Be mindful when stacking. Never stack too high as over-reaching can cause injury. Also, collapse of the stack could cause crush injuries which could be fatal. When laying bricks and blocks, mortar can splash on to skin, causing irritation, or into eyes, and when cutting, hand and eye injury is possible.

> **Definition**
>
> **Kinetic lifting** – a way of lifting objects that reduces the risk of injury to the lifter

Laying bricks

A bricklayer has to lay bricks level and straight with equal sized joints in order to achieve a sound wall and a good overall appearance. Over the next few pages we will look at how this is achieved.

Bonding

Bond is the name given to the pattern of the bricks in a wall. The purpose of bonding brickwork is to distribute the weight of the wall evenly along its length and on to the foundation below. The length of the wall should be set out dry first to see if full size bricks can be used along the length or whether cut bricks will be needed. See Chapter 9 Bonding (page 169).

The purpose of joints

Bricks are often not regular in shape or size (see Figure 10.1) so mortar is used to make up the difference and keep the wall looking neat. By increasing or decreasing the size of the bed joint under each brick, it is possible to keep the top **arris** flat.

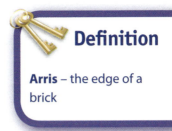

Definition

Arris – the edge of a brick

Figure 10.1 Irregular shaped bricks

By opening up or tightening the **perp** joints between the bricks it is possible to keep the joints neatly above each other. In most cases this is only a fraction as joint sizes should not exceed 10 mm in width.

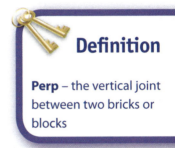

Definition

Perp – the vertical joint between two bricks or blocks

The building process

To build any wall, corners have to be built in the correct positions first (unless using corner profiles). In order to lay the bricks between corners and maintain accuracy, the bricklayer will use a line. The string line is attached level to the course height, and to a pre-erected corner using pins or corner blocks. The line is then pulled tight to take out any 'sag' in the line and attached to the other corner.

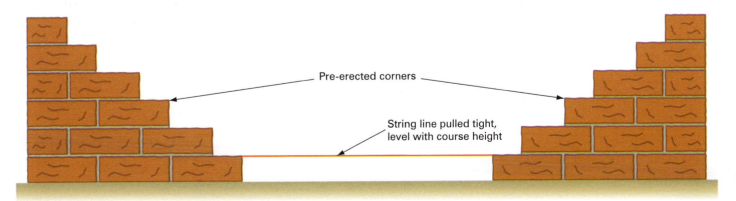

Figure 10.2 Corners with line attached

The bricks can now be laid to the line, which will ensure they are the correct height and are laid in a straight line. The bricks should be tapped down until the top arris is level with the top of the string line.

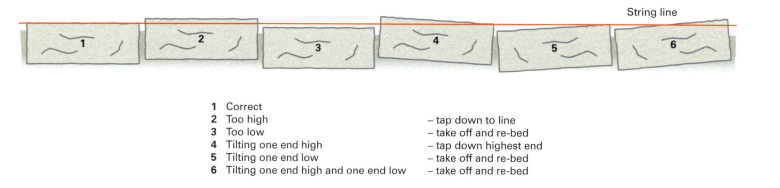

1 Correct
2 Too high — tap down to line
3 Too low — take off and re-bed
4 Tilting one end high — tap down highest end
5 Tilting one end low — take off and re-bed
6 Tilting one end high and one end low — take off and re-bed

Figure 10.3 Bricks laid to a string line

To ensure the bricks are laid in a straight line upwards (face), the bricklayer should look from above the line downwards (see Figure 10.4). There should be a slight gap between the line and the face of the brick. This gap should be about the thickness of a trowel and should be even along the length of the brick. It is important that the brick should not touch the line at any time as this can cause the line to be pushed outwards, resulting in the wall becoming 'curved'. It will also be a nuisance to any other bricklayer who is using the line.

Chapter 10 Laying bricks and blocks to a line

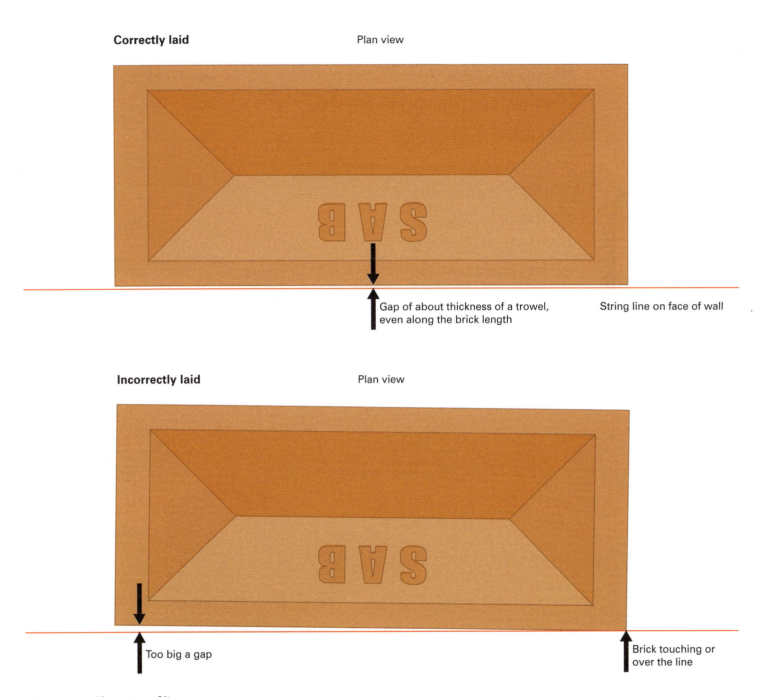

Figure 10.4 Plan view of line

Fixing the lines to the corners

The lines can be attached to the corners by:

1. line pins
2. corner blocks.

183

Brickwork NVQ and Technical Certificate Level 2

Line pins

The line should be wound on to the blade of the pin so that the line is on top of the pin, pointing towards the wall.

The blade is then placed into the perp joint nearest the **quoin**. The pin should be angled slightly downward with the line level to the top of the pre-laid brick.

Definition

Quoin – the corner of a wall

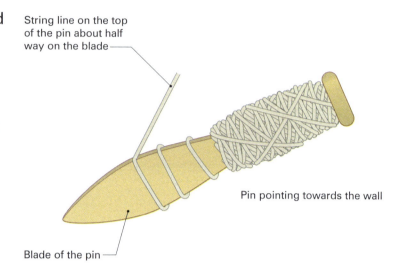

Figure 10.5 Pin

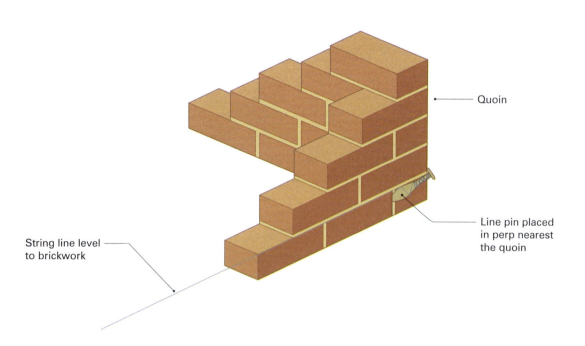

Figure 10.6 Perp joint nearest the quoin with pin attached

Chapter 10 Laying bricks and blocks to a line

Corner blocks

To avoid the 'pin-holes' left in the quoin, the line could be fixed in place using corner blocks. The blocks can be made of wood, plastic or metal (see Figure 10.7). The line is pulled through the corner block and attached to screws or tied.

The corner block is placed on to one quoin with the line pulled tight at all times. The other corner block is then fixed to the line about a ¼ brick short of the quoin, depending on the length of the wall to be built – sometimes it may require more. The block is then pulled and fixed to the second quoin. Both blocks are adjusted so that the line is level with the course of bricks to be laid.

Remember

Be careful not to over tighten as the line can break or the corner blocks could be pulled out of plumb

Note

Corner blocks can only be used on external corners

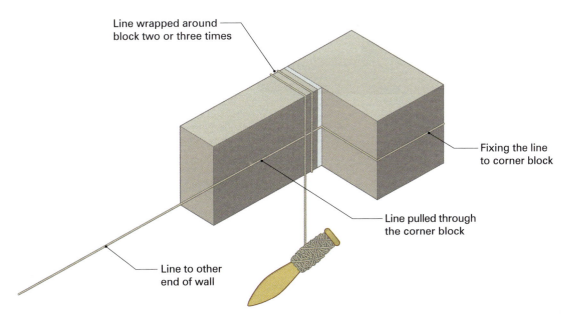

Figure 10.7 Corner block

185

Tingle plate

When building very long courses of brickwork, the line may sag even when pulled tight. If this was not corrected the courses of bricks would dip in the middle. To hold the line up in the middle of the wall a **tingle plate** is used. A tingle plate is a flat piece of steel with three prongs, and the line is fed under the ends and over the middle prong so that the line stays at the bottom.

The tingle plate is placed on top of a previously lined, levelled and gauged brick at the centre of the wall. A brick is then placed on top of the plate to hold it secure, and the course is then laid to the line in the normal way.

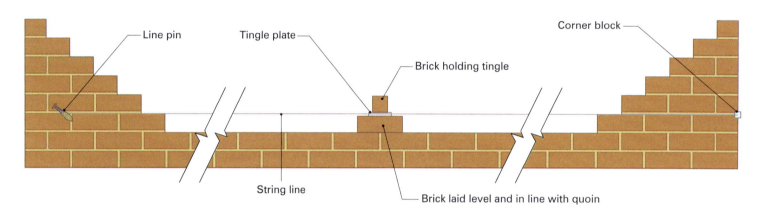

Figure 10.8 Tingle plate on the line

On the job: Building a level wall

Ray is building a brick wall 10 m long. He is using his line and the wall looks nice and straight. When he steps back and looks, there is a sag in the middle. Why has this happened and how can Ray get over this?

Chapter 10 Laying bricks and blocks to a line

Protecting newly laid brickwork

After completing a day's work building a wall, the bricklayer must take precautions to prevent damage to the wall from the weather. Rain will cause mortar joints to run over the face of the wall, causing unsightly stains. Cold weather could cause the water in the mortar to freeze, damaging the bond between the bricks, making the wall weak and possibly causing the wall to be taken down.

To prevent rain damage the walls should be covered with a polythene sheet or tarpaulin. A scaffold board or bricks can be used to secure the cover in place. The sheeting should be kept clear of the wall to allow for ventilation. See Figure 10.9.

Did you know?

To prevent the mortar from freezing, newly-built walls should be covered if there is a sudden drop in temperature below 3°C. The covering should be a roll of hessian or insulation slabs, with a waterproof tarpaulin or plastic sheeting on top

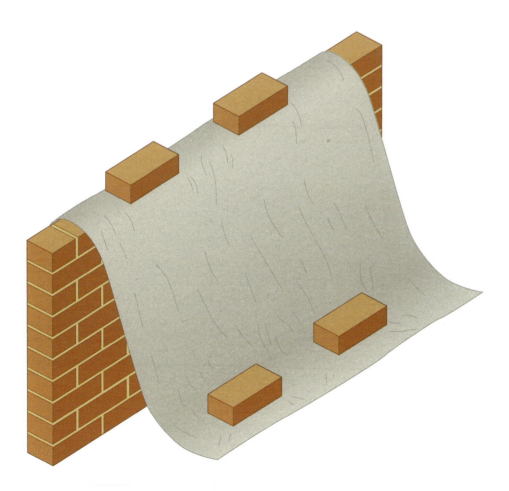

Figure 10.9 Protecting newly laid brickwork

187

Remember

If a wall blows down, it will cost time and money to replace it

Laying blocks

The procedures for setting out and laying blocks are basically the same as for brickwork, but remember that blocks are bigger and heavier, especially concrete blocks. The area of blockwork that can be built during a working day is greater than with brickwork but be very careful about the height that you build – the type of block used will govern this, as well as weather conditions.

Lightweight block walls can usually be built higher as the water content in the mortar is absorbed into the blocks, drying the joints faster and so giving stability more quickly. Concrete blocks are a lot heavier, do not absorb mortar water as quickly and because of this are more likely to compress (swim), with the water inclined to run down the face of the work. Also, because of the weight, the blockwork can start to bow out of plumb.

Blocks should not be laid too high if working in windy conditions as walls can be blown down, causing damage to property or people.

On the job: Covering newly-built walls

Nick is about to build a new dense block wall, 8 m long and 2.1 m high. The blocks have been left uncovered overnight before being loaded out for use. Heavy rain has saturated the blocks. Nick's boss is expecting the wall to be built in a day. Can Nick go ahead and build the wall? What could Nick do with the mortar? If it was a windy day, is there anything else Nick will be unable to do?

Thin block jointing

As the name suggests, this is a system that has bed and vertical joints of approximately 1 to 3 mm in thickness. The mortar used is a cement-based powder that is mixed in proportion with water to the manufacturer's instructions.

The first course of blocks is laid on a traditional sand and cement bed joint of 10 mm and levelled. This course would normally be laid on to a damp proof course. This course is then allowed to dry fully. Once dry it is rechecked for level, with any high spots being shaved off the block. Due to the nature of the joint system, the tolerances are critical. Both the bed and vertical joints are then applied to the blocks by means of a scoop with a serrated edge, to give an even thickness to the joints. The joints will start to set within 10 minutes and reach full strength in 1 to 2 hours.

Did you know?

Because of the setting times using thin block jointing, buildings can be constructed a lot faster than with traditional methods

The blockwork is constructed first up to roof plate level, incorporating first floor joists. The roof trusses can be fixed and the building can be watertight, allowing the internal trades to start before any external brickwork is built. This can cut building time by up to a third.

The brickwork is then built using special cavity ties, which are driven into the blocks – no drilling or fixing is required. Openings can then be cut out to suit requirements, saving time on cutting and plumbing. Corners are still required to be built and great care should be taken to ensure plumb, but remember the setting time is very quick.

The blocks are still built using line and pins as in traditional methods.

Did you know?

Thin block jointing is used in only some areas of the UK. Other countries have been using this system since the 1960s

Brickwork NVQ and Technical Certificate Level 2

FAQ

Can I lay frogged bricks downwards? It's so much easier than laying them with the frog on top.

A frogged brick is one that has an indent or 'depression'. Laying frogged bricks downwards is not a good idea at all as there is no way to ensure that the frog is filled with mortar. This reduces the strength of your wall.

Knowledge check

1. Name the two ways to attach lines for bricklaying.
2. What is used to stop a line from sagging in the middle?
3. What is the main purpose of joints?
4. What can happen if concrete blocks are built too high in one go?
5. Why is a newly-built wall covered at the end of the day?
6. Why are bricks not laid tight against the line?
7. What system uses joints between 1 and 3 mm thick?
8. What is bonding?
9. What might hessian be used for whilst laying bricks?
10. What gap should be left between the line and the brick when laying?

chapter 11
Solid walls

OVERVIEW

Solid walls are walls built entirely from bricks or blocks and mortar with no voids and with little or no other materials used. They range from 102 mm (or ½ brick) thick up to 450 mm (or 2 bricks) thick. In exceptional situations walls may be thicker but this is very rare nowadays – thicker walls would more likely be made of solid concrete.

We will look at the basic principles and uses of the different sizes of walls.

In this chapter we will cover:

- Health and safety
- Types of wall
- Basic rules about bonding and building solid walls
- Reinforcement in brickwork
- Raking cutting
- Joining walls to existing walls
- Vertical movement joints
- Weathering a wall.

Health and safety

As with all materials, great care should be taken with moving, storing and using bricks or blocks as well as the mortar required. Please refer to Chapter 2 Health and safety (page 45) and Chapter 6 Handling and storage of materials (page 122) for more information.

Types of wall

Half brick walls

Half brick walls are 102 mm thick and are nearly always built in stretcher bond or 'half bond' (see Figure 11.1). Half bricks are used on stopped ends to form the corners. Walls should be set out to use as many stretchers as possible and no cuts should be less than a half bat (102 mm).

This type of wall is mainly used as the face wall on cavity walls or garden walls if piers are incorporated, and internal walls in brick or block to partition rooms.

> **Remember**
>
> ½ bats and ¾ bats are used to gain bond

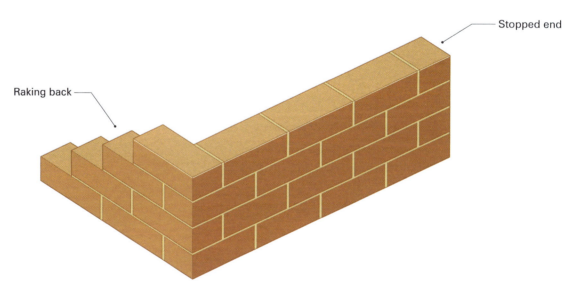

Figure 11.1 Half bond wall

In the case of face walls, the corner pattern should be the same at both ends but is sometimes reversed to help with bonding. If cuts are required, sometimes it is better to place them under a door or window. This should be looked into at the setting out stage but is not always possible.

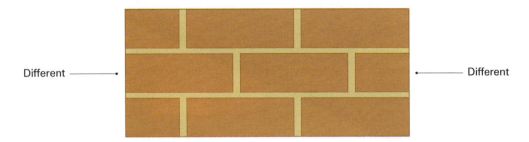

Different — Different

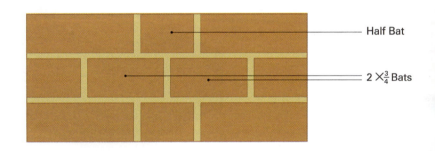

— Half Bat

— 2 × $\frac{3}{4}$ Bats

Result if a reverse bond is not used (broken bond)

Figure 11.2 Reversed bond wall

1 brick walls

1 brick walls are 215 mm thick and enable the bricklayer to place headers across the wall, which make it possible to create several different bonds. This type of wall is used where more strength is required, as with fire walls, garden walls, manholes, areas where steelwork is to sit and for sound proofing. There are four main bonds used:

1. English bond
2. Flemish bond
3. English garden wall bond
4. Flemish garden wall bond.

With these bonds, the bricks lap the course below by a ¼ brick and a queen closer should always be used at the corner (quoin) next to the header.

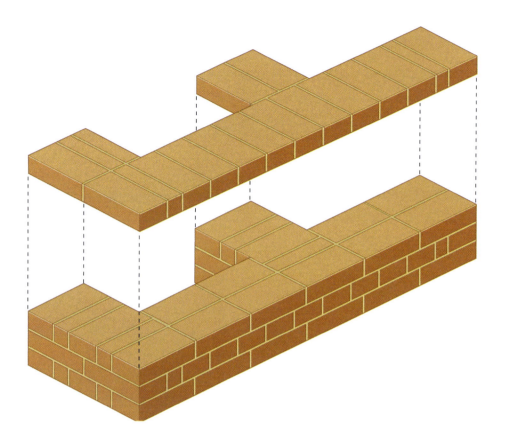

Figure 11.3 1 brick wall

For more information on bonds, see Chapter 9 (page 169).

> **Remember**
>
> Make sure the perps are plumb

On the job: Building a wall in English bond

John has been asked to build a 1 brick wall, 30 m long and 1½ metres high in English bond. The foundation concrete has been put in and all materials are on site ready to use. Any there any problems?

1½ and 2 brick walls

These types of wall are normally used for retaining walls holding back earth or water, load-bearing piers or piers to garden walls. In some instances concrete blocks may be incorporated with brick, with the brick giving the face finish as in the case of garden wall piers that require the strength to hold gates. The centre may be block or even concrete, and retaining walls holding back earth in a tiered garden will look better with a brick finish rather than block – the cost would probably be more if only brickwork was used. The blocks would be laid flat and tied in every three courses.

Did you know?

In some instances wall ties may be used to tie the brickwork to the blocks

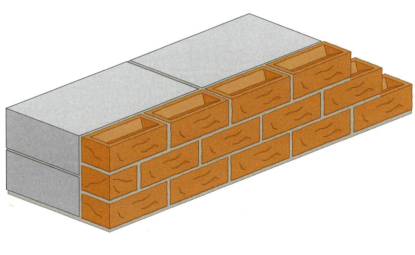

Figure 11.4 1½ brick pier and garden wall with blocks and brick finish

Front Elevation

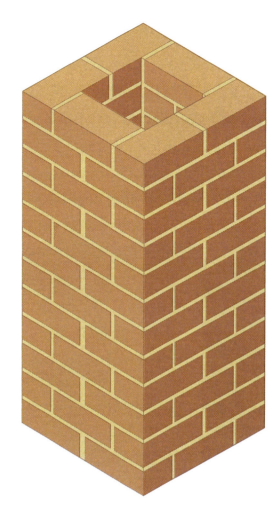

Plan Course 2

Plan Course 1

Figure 11.5 2 brick pier

Basic rules about bonding and building solid walls

- If measurements are given for the length of a wall, set out to avoid unnecessary cutting.
- Avoid broken and reverse bonds if possible.
- Work from the ends to the middle of the wall.
- Keep vertical joints plumb and uniform.
- Avoid wide vertical joints – maximum 10 mm (cannot always be achieved).
- Queen closers should be 46 mm wide to achieve the correct lap.

- Use two ¼ bats instead of a queen closer – it can be easier and prevents waste.
- Take care in English and Flemish bonds to maintain ¼ brick overlap, especially in long walls where you can 'lose' or 'gain' ground.
- Use a line with both sides in garden wall bonds but remember only one side can be plumbed in 1 brick walling.
- Make sure you always lay cut bricks frog up.

Remember

Always set out in stretchers no matter what the bond is

Reinforcement in brickwork

Brickwork is extremely strong in compression (being squashed) but weak in tension (being stretched). If there is any chance of brickwork being subjected to tensile forces (i.e. being stretched) some kind of reinforcement can be built into the wall.

Reinforcing can be done in several ways but the most common way is to build in a steel mesh called 'expanding metal'. The mesh is built into the bed joints. Various widths can be obtained for different thicknesses of wall. Brickwork can also be reinforced vertically with steel rods and concrete.

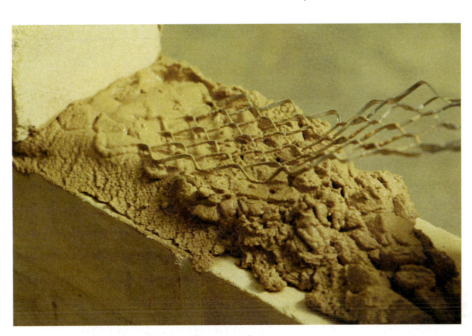

Brickwork reinforced with expanding metal

Brickwork NVQ and Technical Certificate Level 2

Remember

Always check specifications to determine the type and size of reinforcement required

Definition

Subsidence – the sinking in of buildings etc.

Some places where reinforced brickwork could be used are:

- over large span door or window openings
- where high security is required (such as a bank vault)
- in gate pillars
- in foundations where there is a possibility of **subsidence**.
- concrete columns

Column reinforced vertically with steel rods and concrete

Raking cutting

Raking cutting refers to cutting brickwork at an angle, as at the end of a roof.

Raking cutting can be done by hand with the use of a hammer and bolster or by machine using disc cutters or a bench cutting saw.

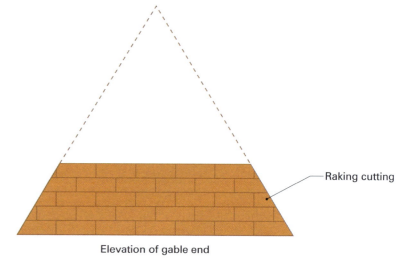

Figure 11.6 Gable end showing raking cutting

Motorised cutting is better where the angle of the cut is very low or the bricks being cut are very hard or contain holes.

Whatever method is used, raking cutting should be neat and a 'guide line' erected before commencement. No brickwork should protrude above the line as it may interfere with whatever is placed on the wall later.

On the job: Cutting bricks for a gable end

Robert has been asked by his boss to finish the brickwork on a gable end. He has discovered that the bricks used have holes in them. He has a hammer and bolster in his tool bag. What should he do?

Joining walls to existing walls

There are different ways to join walls to existing work. The three main ways are described below.

Toothing

Bricks are left out on alternate courses at the end of a wall so that the new wall can be 'toothed' in at a later date. This method is not recommended as it leaves a weakness within the wall because it is very difficult to make the mortar joints solid.

End of wall to be toothed at later date

Indents

Holes are left out in a wall to receive a brick or blockwork junction. Alternate courses can be left out in the case of brickwork but with blockwork, three courses of brickwork are left out to accommodate the blockwork. These are sometimes referred to as block indents.

Proprietary wall connectors

Wall connectors provide a method of joining walls without cutting into an existing wall. A metal plate is attached to the existing wall by plugging and screwing. There are different types of attachment for the new wall. Some have angle brackets that can be attached to the plate by protruding lugs. The brackets are then built into the wall as the work proceeds. Some are part of the main structure and fold out on to the courses, others are slid down the main connector into position, then built in. This method is quicker and less damaging than the other two as little vibration is caused to the wall in the connecting process.

Brickwork with holes for junction

Wall connector with metal plate attached with angle brackets

Remember

When using wall connectors the manufacturer's instructions must be carefully followed

Vertical movement joints

Architects and designers have found through experience that walls that are very long (10 m plus) can crack due to movement caused by:

- temperature change
- wetting and drying of the wall.

To solve the problem of movement, a wall can be divided into smaller lengths by introducing movement joints, which are actually an intended 'straight' joint filled with compressible (squashable) filler. This is then covered with mastic to seal the joint to prevent water penetration but allow movement.

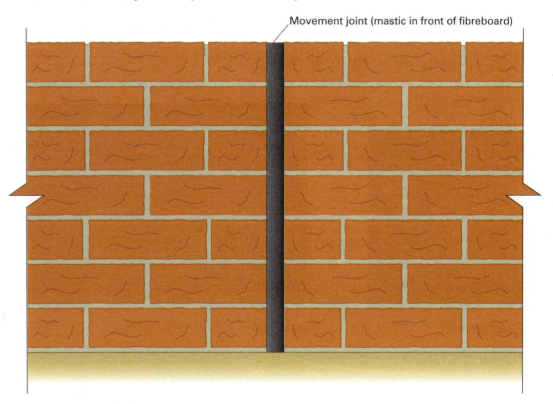

Figure 11.7 Vertical movement joint

Weathering a wall

A protective finish can be added to walls that are likely to be exposed to weather such as rain and frost, which can weaken the wall and cause the top brick course to break up. Cavity walls do not require such protection as they

Definition

Parapet – a low wall that acts as a barrier where there is a sudden drop (e.g. a balcony wall)

are not exposed due to the roof structure covering the top courses. Garden walls, retaining walls and **parapet** walls will require finishing however. A bricklayer can use one of several methods to weather a wall:

- with a brick finish
- with a concrete finish
- with a stone finish.

Brick finish

Common bricks can be used to weather the top of a wall, however, it is advisable to use a hard stock brick or engineering type of brick which are more resistant to water penetration and frost damage. The different finishes that can be achieved will depend on the thickness of the wall. On a ½ brick wall there are two main ways to finish with bricks:

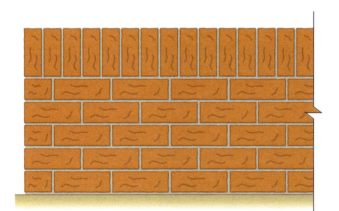

Figure 11.8 A soldier course brick finish

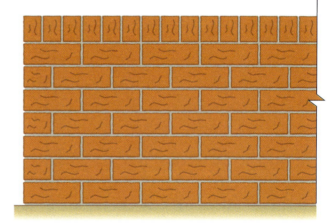

Figure 11.9 A 1/2 bat brick finish

1. The main way to achieve a brick finish is by using a brick-on-end, more commonly known as a soldier course. See Figure 11.8.

2. 1/2 bats can be used in the same way (see Figure 11.9) but they will need to be cut exactly in half, although this does not allow for the difference in brick sizes. In addition, the bricks would have to be cut by machine as cutting by hammer and bolster may not always cut squarely, resulting in variable back joint sizes which gives a poor appearance to the finish.

Chapter 11 Solid walls

On a 1 brick wall, bricks are laid on edge to protect the top (see Figure 11.10). It is always best to use a hard water-resistant type of brick. Sometimes the wall may have what is called a tile creasing under the brick on edge. This is normally two courses of flat concrete tiles bedded on mortar and half bonded. This helps to stop rainwater penetration as water that would normally run down the face of the bricks is pushed away as the tiles are wider than the wall. A brick on edge finish can also be used on walls that are 1½ and 2 bricks wide, as well as a finish to 1, 1½, and 2 brick **piers** (see Figure 11.11).

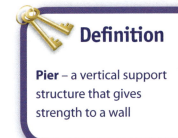

Definition

Pier – a vertical support structure that gives strength to a wall

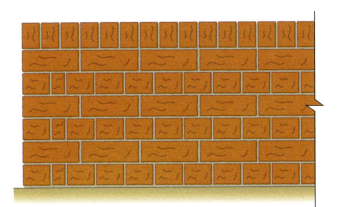

Figure 11.10 A brick on edge finish on 1 brick wall

Figure 11.11 A brick on edge finish on a pier

Concrete finish

Concrete can be used to weather a wall in the form of pre-cast sections called copings. These are usually factory manufactured and come in a range of width sizes to suit requirements but are normally 300 mm or 450 mm in length. There are two main profile types of coping normally used: saddleback and feather edged (see Figures 11.12 and 11.13). Copings are made slightly wider than the wall to allow water to drain past the face of the wall.

Figure 11.12 Saddleback coping   **Figure 11.13** Feather edge coping

Stone finish

Natural stone cut to the desired size can be used to weather a wall in the same way as concrete (see Figure 11.14). Stone will give a more rough-looking finish as not all pieces will be regular sizes. Stone is also a more expensive option for weathering due to the higher cost of the material.

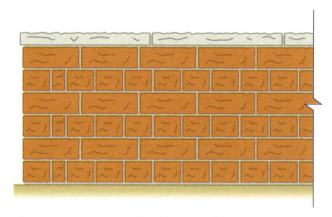

Figure 11.14 A stone finish on a wall

Piers

We have already looked at some of the ways a pier can be weathered but we will now look specifically at caps. A concrete cap can be bedded to the finished brickwork of a pier in order to protect the top (see Figure 11.15). Caps are factory produced and come in a range of standard sizes to suit most piers. The higher centre point causes water to run off the cap and an overhang means that the water drips away from the brickwork, thus offering further protection.

Figure 11.15 Concrete pier cap

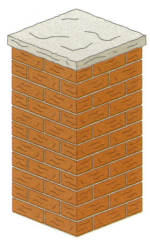

Figure 11.16 Flat stone pier cap

An alternative to a concrete cap is a flat stone cap (see Figure 11.16). As with stone finishes to walls, this choice of finish gives an irregularly-shaped appearance and is more costly than other options.

On some older walls that are 2 or more bricks wide, tiles can be used to form a small roof-like structure to protect the brick or chalk wall that is exposed to the elements.

FAQ

What finish should I use to weather a wall?

You could use a soldier course for a 1/2 brick wall or a brick on edge for a 1 brick wall.

Knowledge check

1. Name the four main bonds used for a 1 brick wall.
2. What would a 1½ brick wall be used for?
3. Name the three main ways of joining a new wall to an existing wall.
4. What is the main rule on raking cut work?
5. What bond is mostly used for ½ brick walls?
6. If brickwork is subjected to tensile force what could be added to help?
7. What should always be used next to the corner header on a 1 brick wall?
8. On long walls, what is put in to stop cracking?
9. How wide should a queen closer be?
10. What is the bond called when the corners start with opposite bricks?

chapter 12
Cavity walling

OVERVIEW

Cavity walls are mainly used for house building and extension work to existing homes and flats. They consist of two separate walls built with a cavity between, joined together by metal ties. In most cases the outer wall is made of brick with the inner skin made of block.

The main reason for this type of construction is to protect the inside from water penetration. The cavity forms a barrier: as the outer wall becomes wet, through the elements, water is not passed through as the two walls do not touch. Air circulates around the cavity to dry the dampness caused, as well as keeping the inner wall dry. In positions that the walls do meet, for example at door and window openings, a damp proof course (DPC) is used to stop water penetration. The cavity can be insulated either partially or fully to make the building warmer and energy-efficient.

In this chapter we will look at:

- How cavity walls are constructed
- Types of insulation
- Openings in cavity walls.

How cavity walls are constructed

Cavity walls mainly consist of a brick outer skin and a blockwork inner skin. There are instances where the outer skin may be made of block and then rendered or covered by tile hanging. The minimum cavity size allowed is 50 mm but the cavity size is normally governed by the type and thickness of insulation to be used and whether the cavity is to be fully filled or partially filled with insulation.

The thickness of blocks used will also govern the overall size of the cavity wall. On older properties, the internal blocks were always of 100 mm thickness. Nowadays, due to the emphasis on energy conservation and efficiency, blocks are more likely to be 125 mm or more.

In all cases, the cavity size will be set out to the drawing with overall measurements specified by the architect and to local authority requirements.

Once the **foundations** have been concreted the **footings** can be constructed, usually by using blocks for both walls (see Figure 12.1).

In some situations trench blocks may be used below ground level and then traditional cavity work constructed up to the damp proof course (DPC). A horizontal DPC must be inserted at a minimum height of 150 mm above

Definition

Foundations – concrete bases supporting walls

Footings – brickwork between the foundation concrete and the horizontal damp proof course (DPC)

Remember

The correct size must be used for the internal wall, with the cavity size to suit

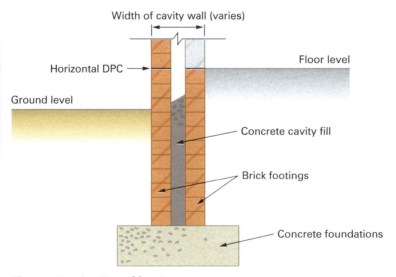

Figure 12.1 Section of footings

Chapter 12 Cavity walling

ground level to both walls. This is to prevent damp rising, below ground, up through the block and brickwork to penetrate to the inside. The cavity must also be filled with weak concrete to ground level to help the footing resist the pressure of the soil against the external wall and oversite fill material.

Damp proof course (DPC)

Damp proof course (often shortened to DPC) is a layer of non-absorbent material bedded on to a wall to prevent moisture penetrating into a building. There are three main ways moisture can penetrate into a building:

1. rising up from the ground

2. through the walls

3. moisture running downwards from the top of walls around openings or chimneys.

There are three types of DPC:

1. flexible

2. semi-rigid

3. rigid.

Flexible DPC

Flexible DPC comes in rolls of various widths to suit requirements. Nowadays most rolls are made of pitch-polymer or polythene but bitumen can still be found. Metal can be used as a DPC (in copper and lead) but because of the cost is mainly used in specialised areas. The most widely used and economic DPC material is polythene. Flexible DPC should always be laid upon a thin bed of mortar and lapped by a minimum of 100 mm on a corner or if joining a new roll.

Semi-rigid DPC

This type of DPC is normally made from blocks of asphalt melted and spread in coats to form a continuous membrane for tanking for basements or underground work.

Rigid DPC

Rigid DPC uses solid material such as engineering bricks or slate, which were the traditional materials used. Slate is more expensive to use than other DPC materials and has no flexibility. If movement occurs the slate would crack, allowing damp to penetrate. Engineering bricks could be used for a garden wall if a DPC was required.

Floors

There are two main types of floor used for most modern day houses: solid floor construction and suspended concrete flooring.

Solid floor

A solid floor consists of a hardcore base of a porous material with a minimum depth of 150 mm on which a solid concrete slab is laid. The concrete should have a minimum thickness of 100 mm. A thin layer of sand should be laid on top of the hardcore to fill any voids; this is usually called sand 'blinding' on to which a polythene membrane is laid to stop any dampness from rising up through the ground and making the concrete slab (floor) wet.

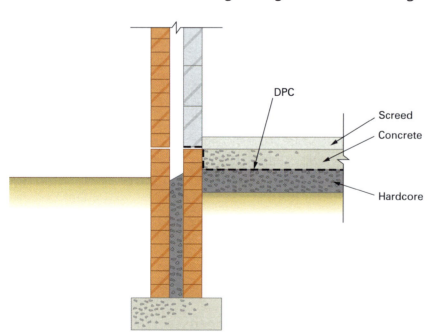

The membrane should be returned up the wall so as to return under the DPC on top of the wall. The slab should finish at the same height as the DPC level, so existing ground should be dug out to suit the correct levels to accommodate the hardcore etc. The floor is normally finished with a 50 mm cement and grit screed.

Figure 12.2 Solid floor construction

Suspended concrete floor

A suspended concrete floor consists of reinforced concrete beams that lay on to the inner leaf of the blockwork with standard dense concrete blocks laid between to form the floor. There should be a minimum depth of 150 mm between the bottom of the floor and ground to allow air circulation and to stop damp from rising.

Air bricks must be situated in the outer leaf below DPC level, to allow a free flow of fresh air, and **cavity liners** inserted into the inner leaf connected to the airbricks. A DPC tray must be situated above the liner to prevent damp penetration to the inner leaf, usually caused by dropped mortar being left on top of the liner (see Figure 12.4).

The floor can be finished with a 75 mm reinforced screed or timber flooring laid on a polythene membrane.

Definition

Air bricks – bricks with holes to allow air to pass through a wall

Cavity liner – placed behind an air brick to form an air duct

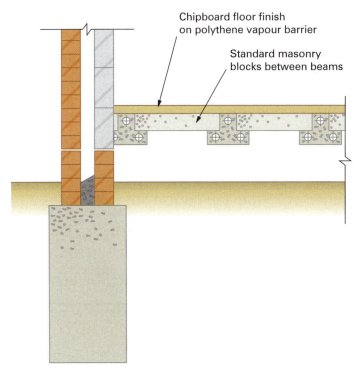

Figure 12.3 Suspended floor construction

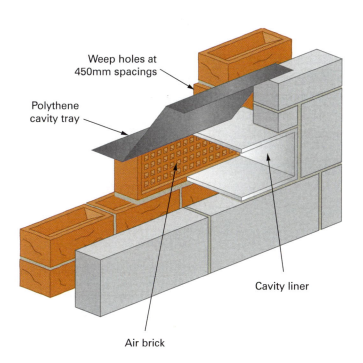

Figure 12.4 Airbricks with cavity liner and tray

On the job: Building a cavity wall up to DPC level

Kevin works for a medium-size construction company on the sheltered housing building section. The work has been completed up to DPC level by groundwork contractors, using a beam and block floor system. The external brickwork is continuous up to DPC in 105 mm face bricks. Is this correct?

Cavity walls above DPC

The older, traditional way to build a cavity wall is to build the brickwork first and then the blockwork. Now, due to the introduction of insulation into the cavity, the blockwork is generally built first, especially when the cavity is partially filled with insulation. This is because the insulation requires holding in place against the internal block wall, by means of special clips that are attached to the **wall ties**. In most cases the clips are made of plastic as they do not rust or rot. The reason for clipping the insulation is to stop it from moving away from the blocks, which would cause the loss of warmth to the interior of the building, as well as causing a possible **bridge** of the cavity, which could cause a damp problem.

The brick courses should be gauged at 75 mm per course but sometimes course sizes may change slightly to accommodate window or door heights. In most instances these positions and measurements are designed to work to the standard gauge size. This will also allow the blockwork to run level at every third course of brick, although the main reason will be explained in the wall tie section on the next page.

On most large sites, corner profiles are used rather than building traditional corners (see Figure 12.5). These allow the brickwork to be built faster and, if set up correctly, more accurately. But they must also be marked for the

> **Definition**
>
> **Wall ties** – metal or plastic fixings to tie cavity walls together
>
> **Bridge** – where moisture can be transferred from the outer wall to the inner leaf by material touching both walls

Chapter 12 Cavity walling

gauge accurately and it makes sense to mark window sill heights or **window heads** and door heights so they do not get missed, which would result in brickwork being taken down.

Definition

Window head – top of a window

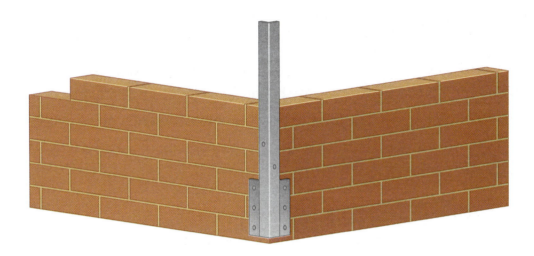

Figure 12.5 A corner profile set up

Wall ties

Wall ties are a very important part of a cavity wall as they tie the internal and external walls together, resulting in a stronger job. If we built cavity walls to any great height without connecting them together, the walls would be very unstable and could possibly collapse.

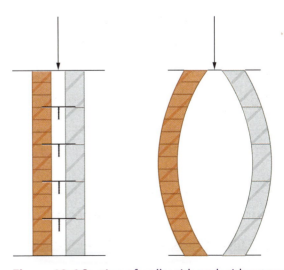

Figure 12.6 Section of walls with and without wall ties

A wall tie should be:

1. rust-proof
2. rot-proof
3. of sufficient strength
4. able to resist moisture.

The most common type is known as the butterfly wall tie. It is made out of strong stainless steel wire in the form of two triangles. The ends of the wire are twisted together to form a drip to prevent the passage of water being carried from the outer wall to the inside wall.

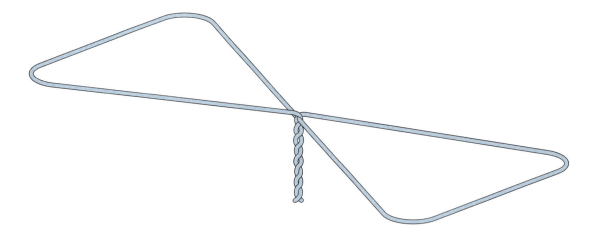

Figure 12.7 A butterfly wall tie

Care must be taken to keep the wall ties clean when placed in the wall because if bridging occurs it may result in moisture penetrating the internal wall.

The positioning of wall ties is extremely important to obtain stability, and the spacing of the ties must be in accordance with the Building Regulations. Figure 12.8 shows the general pattern of wall ties, in the inner block wall, looking from the inside of the cavity.

Chapter 12 Cavity walling

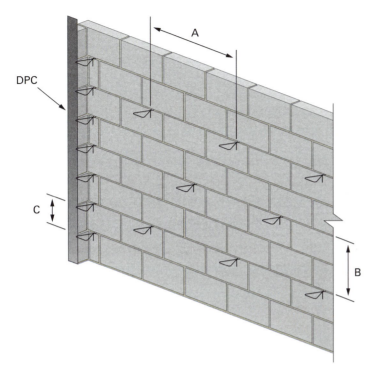

Figure 12.8 Ties in block wall

A = The maximum horizontal distance: 900 mm

B = The maximum vertical distance: 450 mm

C = The maximum vertical distance: 300 mm

Wall ties are set in a staggered pattern through the length of the wall. The maximum vertical distance at C is situated at a window or door reveal. 300 mm is the maximum distance apart, but to work to brick and block courses this measurement is normally 225 mm.

Keeping a cavity wall clean

It is important to keep the cavity clean to prevent dampness. If mortar is allowed to fall to the bottom of the cavity it can build up and allow the damp to cross and enter the building. Mortar can also become lodged on the wall ties and create a bridge for moisture to cross. We can prevent this by the use of **cavity battens**, pieces of timber the thickness of the cavity laid on to the wall ties and attached by wires or string (to prevent dropping down the cavity) to the wall and lifted alternately as the wall progresses.

Did you know?

Any batten can be used as long as the width is the same as the cavity space

Definition

Cavity batten – a timber piece laid in a cavity to prevent mortar droppings falling down the cavity

215

Brickwork NVQ and Technical Certificate Level 2

Remember

Clean at the end of each day, as the mortar will go hard. Good practice is to lay hessian across the wall ties and remove it at the end of each day to clear any mortar that has dropped down the cavity

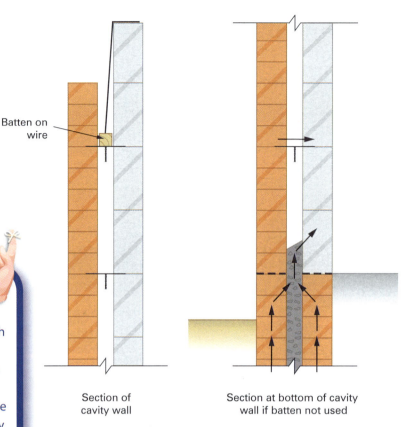

Figure 12.9 Cavity batten in use

The bottom of the wall can be kept clean by either leaving bricks out or bedding bricks with sand so they can be taken out to clean the cavity. These are called core holes and are situated every fourth brick along the wall to make it easy to clean out each day. Once the wall is completed the bricks are bedded into place.

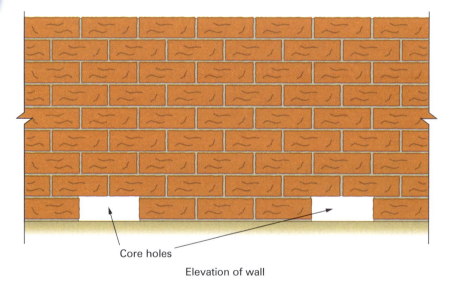

Figure 12.10 Core holes

216

Types of insulation

Cavity walls are insulated mainly to prevent heat loss and therefore save energy. The Building Regulations tell us how much insulation is required in various situations, and in most cases this would be stipulated in the specification for the relevant project to obtain planning permission from the local council.

Cavity insulation can be either Rockwall or polystyrene beads.

There are three main ways to insulate the cavity:

1. Total or full fill.
2. Partial fill.
3. Injection (after construction).

Total or full fill

Figure 12.11 shows a section of a total fill cavity wall. The cavity is completely filled with insulation 'batts' as the work proceeds. The batts are 450 mm x 1200 mm, are made of mineral fibres, and placed between the horizontal wall ties.

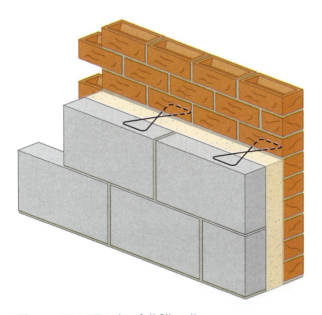

Figure 12.11 Total or full fill wall

Partial fill

Figure 12.12 shows a partial filled cavity, where the cavity insulation batts are positioned against the inner leaf and held in place by a plastic clip. More wall ties than usual are used to secure the insulation in place.

> **Remember**
>
> When injecting insulation, great care must be taken not to drill the bricks as they will be difficult and costly to replace

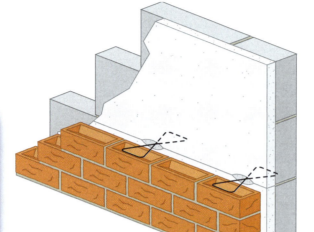

Figure 12.12 Wall with partial fill cavity

Injection

This is where the insulation is injected into the cavity after the main structure of the building is complete. Holes are drilled into the inner walls at about 1 m centres and the insulation is pumped into the cavity. The two main materials used are Rockwool fibreglass or polystyrene granules. The holes are then filled with mortar. If an older property were injected, then the holes would be drilled into the external mortar joints.

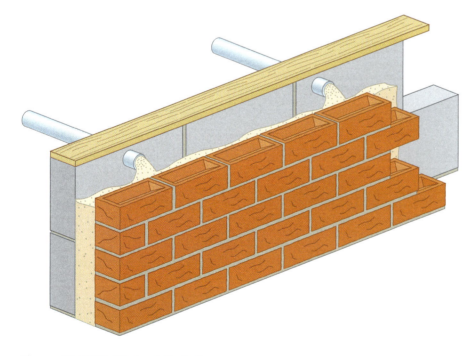

Figure 12.13 Wall being injected

There are three key points regarding insulating cavity walls:

1. Handle and store insulation material carefully to avoid damage or puncturing.
2. Cavities should be clean.
3. Read drawing specifications and follow manufacturers' instructions carefully.

Openings in cavity walls

Openings are put in to walls in the form of door and window frames to allow entry and give natural light to a property. The frames can be made of wood, metal or UPVC. Whichever type of frame is used, the frame has to be secured in place.

Did you know?

The number of frames and their sizes are calculated according to the amount of light required for a room to meet Building Regulations

Wooden frames

Wooden frames are bedded on to the wall using mortar with the brickwork, then built to the sides. Ties should be used to fix the frame to the brickwork to stop the frame from moving or even falling out. There are many different types of ties that can be used in this process. During construction of the wall, the frame should be held in place normally using timber or scaffold boards to stabilise it until the mortar has set. The back edge of the frame should sit flush with the back edge of the brick to help prevent dampness.

Metal frames

Metal frames usually come in either galvanised steel or aluminium and can either be built into the brickwork in the same way as wooden frames or incorporated in to a wooden frame. If they are put in without a wooden frame the sill is usually made of preformed reinforced concrete. Special ties are used to secure to the brickwork.

UPVC frames

These are the most popular type used nowadays as very little maintenance is required after fitting. Made of reinforced plastic, this type of frame is fitted into completed brickwork, therefore the opening left has to be very accurate. The opening is usually formed by using 'dummy frames', which is timber fixed to make a temporary frame 10 mm bigger than the size required. This allows expansion of the plastic once fitted. They are normally fitted by drilling through the frames into the finished brickwork, and then secured by means of plugs and screws. The gap left on the outside is then filled with expanding mastic to prevent water penetration but allow expansion.

On the job: Building a wall that is to have UPVC windows

Ayesha is building cavity walling for an extension. She has built the wall up to window sill height but no windows are on site. She has been told that UPVC windows are to be used. What information does she require to carry on with the extension?

Damp penetration

Where openings are found in cavity walls, much care must be taken to prevent water or dampness from entering the building. The main areas of penetration are shown in Figure 12.14.

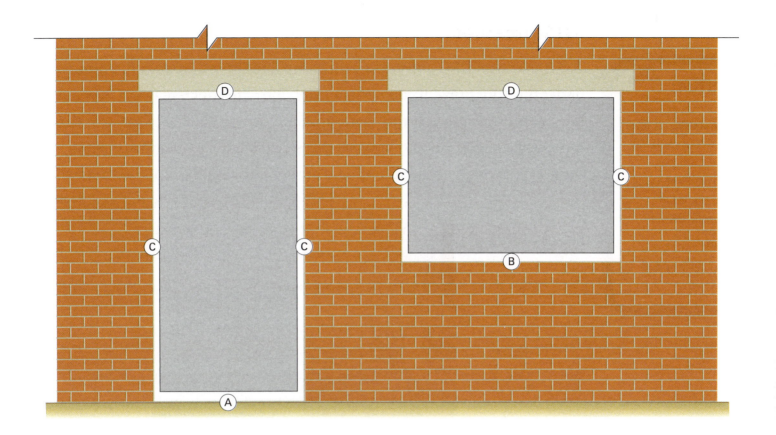

Figure 12.14 Damp penetration areas

A = The threshold of the doorway where rising damp from the floor can be a problem and the door itself can collect rainwater which runs down and collects at this point

B = The sill of the window, which has to deal with any water that runs off the window itself

C = The jambs area of the openings, the main problem here being the passage of moisture between the outer and inner skin of the cavity wall

D = The head of the opening where a structural support is required. Any water entering the cavity above the head must be prevented from sitting on top of the head and channelled to the outside

We will now break down these areas and look at them in more detail.

At the threshold

The **horizontal DPC** prevents rising damp and the waterproof membrane under the floor prevents dampness rising into the concrete floor. The **cavity fill** slopes outwards to throw water that runs down the cavity towards the outer leaf, and the sloping door sill throws water away from the door. All these points help prevent dampness entering the building at the threshold.

> **Definition**
>
> **Horizonal DPC** – a layer of impervious material (water will not pass through) built in a wall 150 mm minimum above ground level
>
> **Cavity fill** – concrete put in a cavity up to ground level

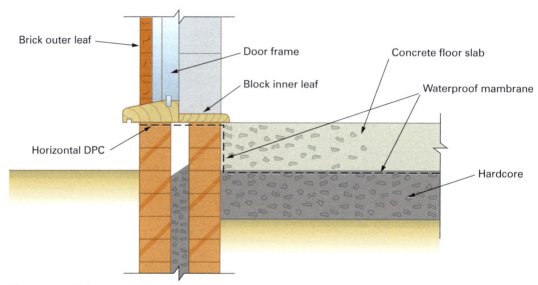

Figure 12.15 Section at threshold

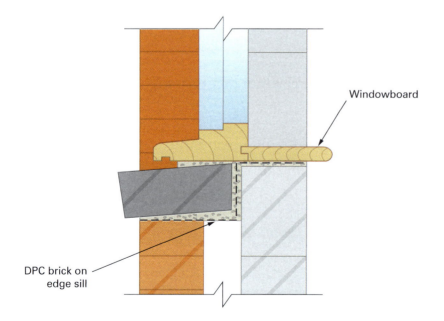

Figure 12.16 Section at sill level

At window sill level

Figure 12.16 shows a groove under the window sill. This is called the **throating** or **drip** and prevents water from running across the bottom of the sill. The brick on edge is sloping outwards to throw water outwards. It is bedded on a DPC to prevent the passage of dampness/water at this point.

At jambs

At the jambs (sides of frames) the penetration of dampness is prevented by a **vertical DPC** (VDPC) as shown in Figure 12.17. It is placed at the back of the outer leaf and must be carefully positioned to be effective.

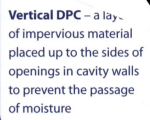

Vertical DPC – a layer of impervious material placed up to the sides of openings in cavity walls to prevent the passage of moisture

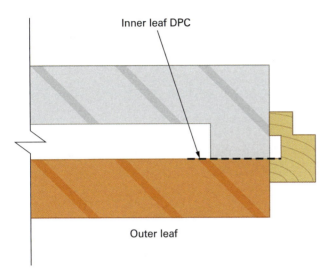

Figure 12.17 Jambs

At head level

The top or 'head' of the frame has to be supported to carry the weight of brickwork and blockwork above this height. Therefore a lintel has to be positioned. The two main types of lintel used are usually 'Catnic' or 'IG', and in both instances a DPC tray must be used above to prevent water/moisture penetration.

With the 'Catnic', the central frame of the lintel slopes outwards to throw any water towards the outer leaf where it is discharged through **weep holes** in the outer wall. The ends of the tray must be turned up (enveloped) to prevent the water running off the end and back down into the cavity. Both types of lintel are made of steel and must have a minimum end bearing of 150 mm and be the correct type to accommodate the width of the cavity used as well as the thickness of block.

Definition

Weep holes – vertical joints left out in brickwork for water to run out

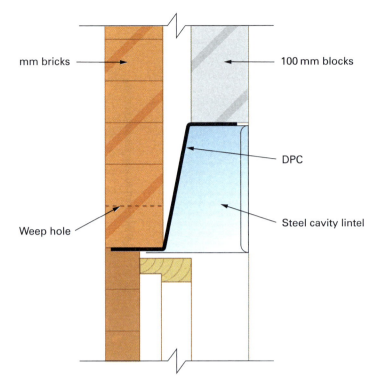

Figure 12.18 Head of frame

Manufacturers' catalogues show the ranges available with widths and lengths for easy ordering but the drawing specification will usually show the type required.

Steps to take to prevent damp penetration

- Set out openings carefully to avoid awkward bonds.
- Care is needed in construction to make sure dampness or water does not enter the building.
- DPCs and wall ties should be carefully positioned.
- Steel cavity lintels should have minimum 150 mm bearings solidly bedded in the correct position.
- Weep holes should be put in at 450 mm centres immediately above the lintel in the outer leaf.

No insulation has been shown in the drawings because they only show one situation. In most cavity wall construction, insulation of one kind or another will have to be incorporated to satisfy current Building Regulations.

Closing at eaves level

The cavity walls have to be 'closed off' at roof level for two main reasons:

1. To prevent heat loss and the spread of fire.
2. To prevent birds or vermin entering and nesting.

> **Remember**
>
> You must read your drawings and specifications carefully to see what is required and always fix insulation to manufacturers' instructions

This area of the wall is where the roof is connected, by means of a timber wall plate bedded on to the inner leaf. The plate is then secured by means of restraint straps that are galvanised 'L' shaped straps screwed to the top of the wall plate and down the blockwork. This holds the roof structure firmly in place and also prevents the roof from spreading under the weight of the tiles etc. The minimum distance that the straps should be apart is 1.5 m. In some instances they may be connected directly from the roof truss to the wall.

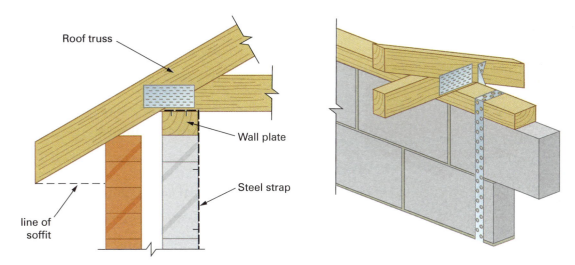

Figure 12.19 Roof section

If a gable wall is required, restraint straps should be used to secure the roof to the end wall (see Figure 12.20).

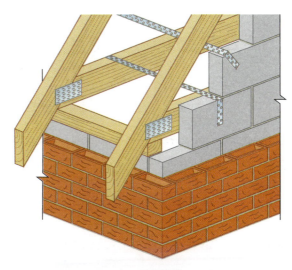

Figure 12.20 Roof section showing gable restraints

The external wall can be built to the height of the top of the truss so as not to leave gaps, or 'closed off' by building blocks laid flat to cover the cavity above the external soffit line from inside, avoiding damp penetration. In some instances the cavity may be left open with the cavity insulation used as the seal.

Gable ends

A gable end is where the wall carries on above eaves level to 'fill in' the shape of the roof usually in the form of a triangle. Figure 12.21 shows the different parts of a roof with a gable end.

When constructing gable ends there are three main considerations:

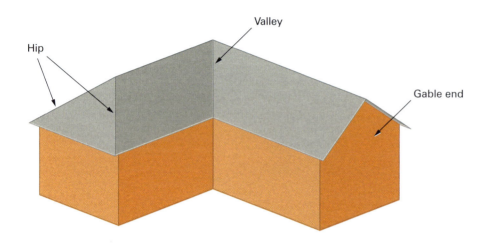

Figure 12.21 Roof with gable

1. How do we start at the eaves?
2. How do we maintain accuracy of the raking cut?
3. How do we maintain plumb?

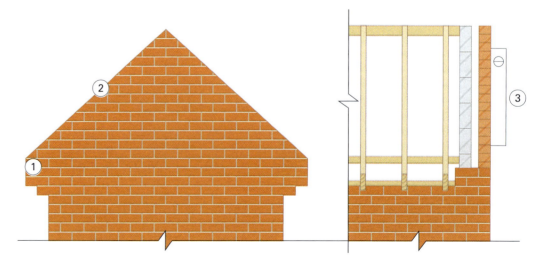

Figure 12.22 Three main considerations when constructing gable ends

Chapter 12 Cavity walling

How do we start at the eaves?

Gable ends very often start with projected bricks or **corbels**; they fill in the void that is made by the eaves of the roof. Other materials can be used at this point such as concrete or tiles.

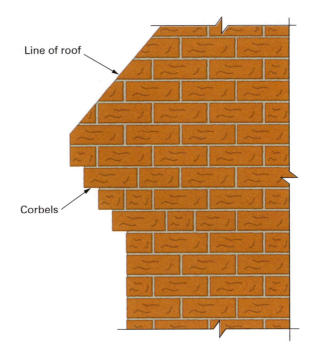

Figure 12.23 Corbels

Definition

Corbel – several courses of bricks laid in front or behind the normal face line to produce a feature on a gable end

How do we maintain accuracy of the raking cut?

Timber laths or battens are fixed to the roof trusses or rafters. The wall is plumbed up and marked on the top lath(s). A string line is then pulled between the laths to maintain the accuracy of the cutting line and maintain plumb.

Did you know?

The gable may have a timber barge board with soffit so cutting is not so critical as long as it is not above the roof line

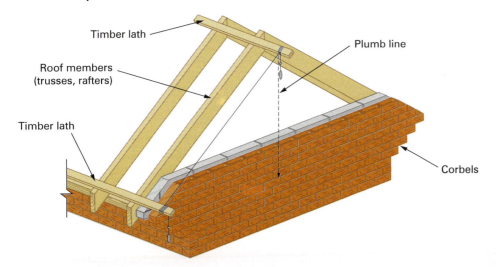

Figure 12.24 A plumb line should be used to maintain accuracy

Brickwork NVQ and Technical Certificate Level 2

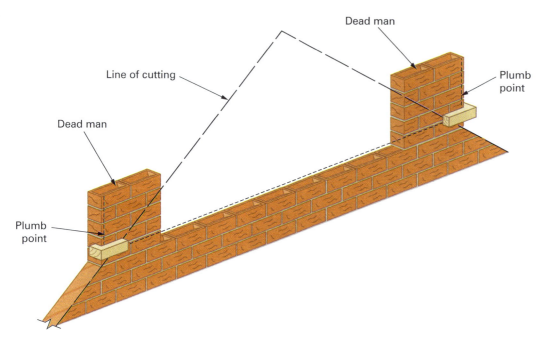

Figure 12.25 Dead men

How do we maintain plumb?

When constructing raking cuts on gable ends the plumb point disappears as you progress so temporary plumb points called **dead men** can be used. These can be formed either by building bricks plumb or by fixing timbers and pulling a string line between them.

Knowledge check

1. What should always be put in above a frame to stop damp penetration into a building?
2. What is horizontal DPC used for?
3. What is the 'head' of the frame?
4. What are the three main methods of insulation?
5. Name the three main types of door and window frame used.
6. What can be used instead of building external corners?
7. What is used for access at the bottom of a wall to clean the cavity?
8. What are used to stabilise the inner and outer leaves of a cavity wall?
9. What is the minimum distance allowed between ground level and DPC?
10. What does the abbreviation VDPC stand for?

chapter 13

Arch construction

OVERVIEW

There are many different arch types and designs. However, their main purpose is the same and that is to provide both an effective and decorative means of bridging an opening. This chapter will deal with the more common, basic types of arch and the methods used for their construction. For information purposes, some of the more detailed and complex arch types and designs are shown at the end of this chapter.

This chapter will cover:

- Parts of the arch
- Geometry
- Methods of constructing simple arches
- Temporary support
- Complex arch types.

Parts of the arch

Arches have been used for many centuries within very different types of construction including bridges, viaducts, aqueducts, castles and the simplest and most modern forms of housing structures.

The curved shape of an arch allows the weight of the masonry above the arch (the load) to be distributed evenly down through the walls at either side. The design means that the arch is not exposed to tensile stresses (pulling apart forces) as it is wedged between the walls on either side and will not collapse when a load is placed above it.

There are certain terms used when referring to parts of an arch, particularly during its construction, all of which you will come across in this chapter. Figure 13.1 shows many of the different parts of an arch and Table 13.1 describes what these are.

Definition

Abutments – the walls or structures through which the weight above the arch is distributed; also, the walls or structures supporting the ends of the arch

Letter	Part	Letter	Part
A	Span – the distance between the **abutments** that support the arch	G	Springer – the first brick of the arch seated on the springing line
B	Springing line – the line at which the arch sits on the abutments	H	Key (or key brick) – the central brick or stone at the top of an arch
C	Rise – the height of the arch from the springing line to the soffit	I	Skewback – the angle at the springing point, on the abutments, at which the arch ring bricks will be laid
D	Intrados – the interior and lower line or curve of the arch	J	Collar joint – the horizontal or bed joint separating the arch rings
E	Extrados – the outside line of the arch ring	K	Crown – the very top point of the extrados
F	Springing point – the point at which the arch meets the abutments	L	Soffit – the underside face of the arch

Table 13.1 Arch construction terminology

Chapter 13 Arch construction

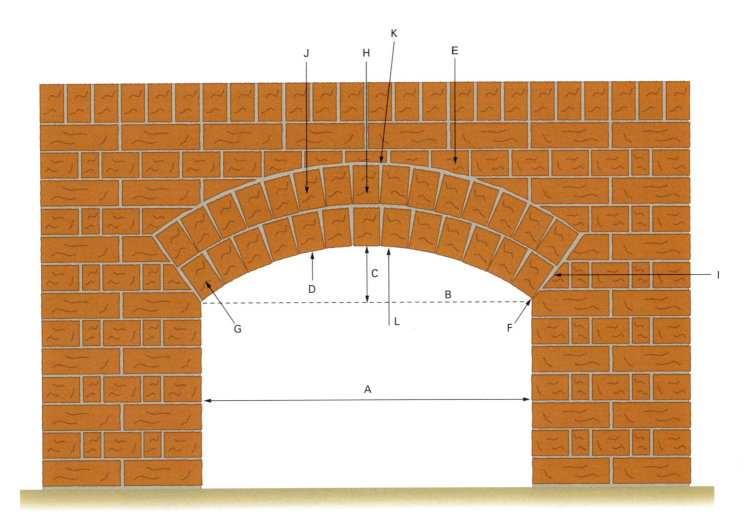

Figure 13.1 The parts of an arch

Some other important arch construction terminology includes:

- Haunch – the bottom part of the arch ring from the springing point to half way up the ring.
- Radius – the distance from the central point on the springing line (known as the striking point) to the intrados.
- Striking point – the central point of the springing line from which the arch radius is struck.
- Voussoirs – the wedge shaped bricks/stones of which the arch is made.

231

Geometry

Geometry is the part of mathematics that deals with lines, points, surfaces and curves. A basic knowledge of geometry is essential when constructing an arch and/or its temporary support (turning piece). Figure 13.2 identifies some of the basic geometrical terms you need to become familiar with.

Figure 13.3 shows some basic setting out rules for bisecting lines and angles.

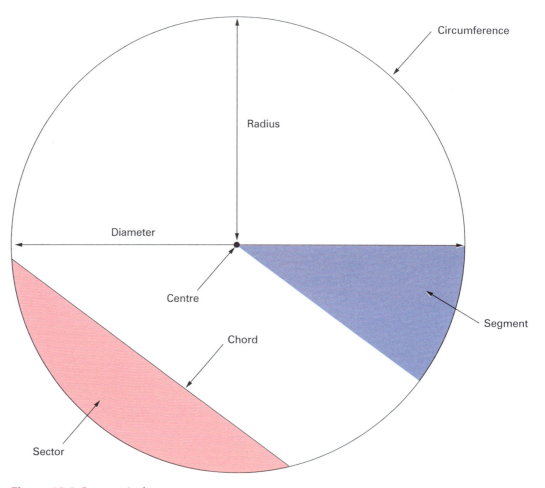

Figure 13.2 Geometrical terms

Chapter 13 Arch construction

Bisecting a line

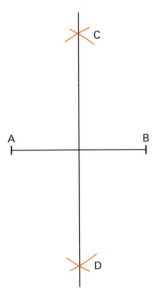

Place compasses on 'A' and 'B' and strike arcs at 'C' and 'D'

Erecting a perpendicular

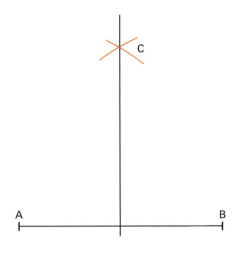

Place compasses on 'A' and 'B' and strike arcs at 'C'

Bisecting an angle

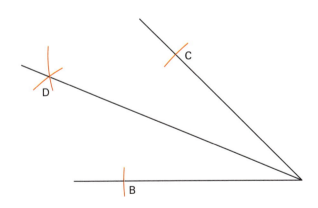

Place compasses on 'A' and mark off equal distances at 'C' and 'D'

With compasses on 'C' and 'B' strike arcs at 'D'

Determining the centre point of an arc

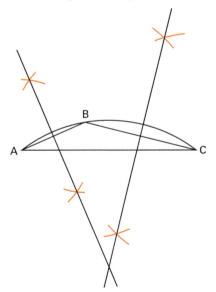

Bisect lines 'A' – 'B' and 'B' – 'C'

Where the bisectors intersect will be the centre of the arc

Figure 13.3 Basic geometrical setting out

233

Setting out a segmental arch

As Figure 13.9 shows on page 236, a segmental arch is one where the intrados, although circular in shape, is less than a semi-circle. We will now look at how you would go about setting out a segmental arch using the geometrical methods we have just covered.

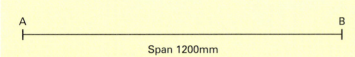

Figure 13.4 Establish the span (a length of 1200 mm has been used here, shown as A–B)

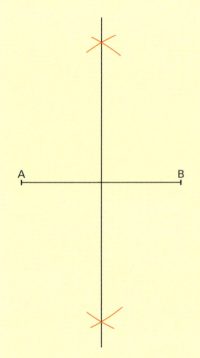

Figure 13.5 Bisect this line (look back at Figure 13.3 page 233 for how to do this)

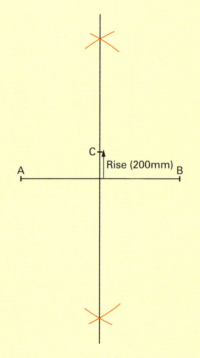

Figure 13.6 Establish the rise (the distance from the springing line (A–B) to the highest point of the soffit shown as C). The rise is normally one sixth of the span, so in this case, the rise is shown as 200 mm

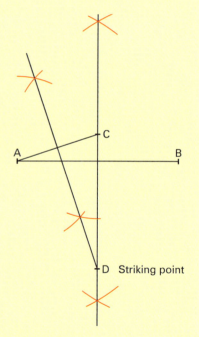

Figure 13.7 Draw a line from A to C and bisect this line. The point where this bisecting line crosses the bisecting line of the span will be the striking point for the arch (shown here as point D)

Chapter 13 Arch construction

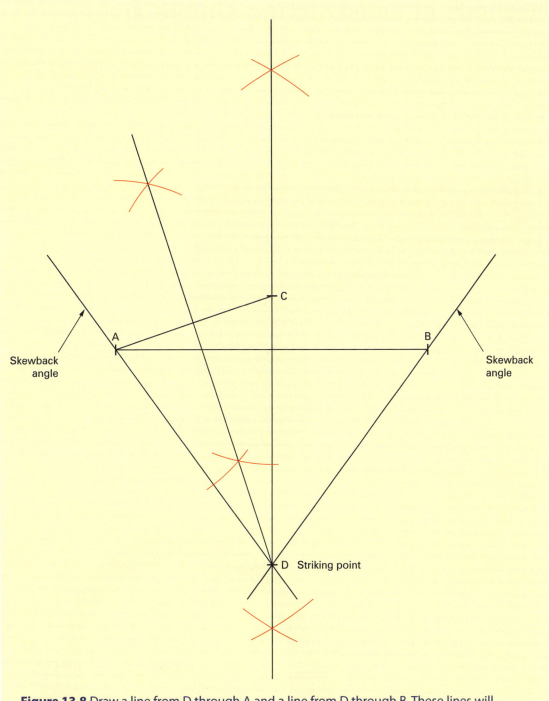

Figure 13.8 Draw a line from D through A and a line from D through B. These lines will provide the angle for the skewbacks.

Methods of constructing simple arches

Two of the most common shapes of arch are the segmental (Figure 13.9) and semi-circular (Figure 13.10). Both the segmental and semi-circular arches can be constructed by using either the rough ringed arch or axed arch methods, both of which we will look at next.

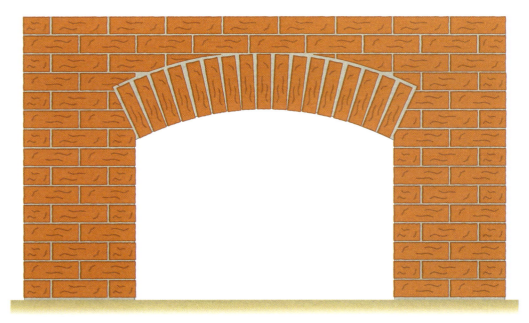

Figure 13.9 An example of a segmental arch

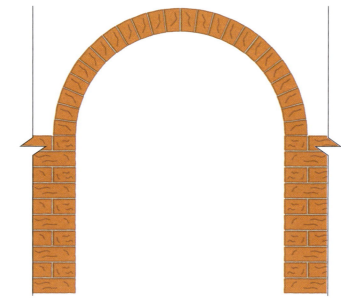

Figure 13.10 An example of a semi-circular arch

Chapter 13 Arch construction

Rough ringed arch

This method consists of using wedge shaped joints with standard sized bricks to form the arch ring. The size of the wedge shaped joints are determined by the **arch centre** or **turning piece**. The bricks that are used in the arch ring are normally laid as headers as opposed to stretchers (see Figure 13.11). The overall height of the arch would then be reached by using a number of arch rings. This is due to the fact that if stretchers were used, the joints would need to be much wider to ensure that the desired shape was obtained. This could result in an unsightly appearance.

Definition

Arch centre – the temporary timber support for the arch ring on semi-circular arches during construction

Turning piece – the temporary timber support for the arch ring on segmental arches during construction

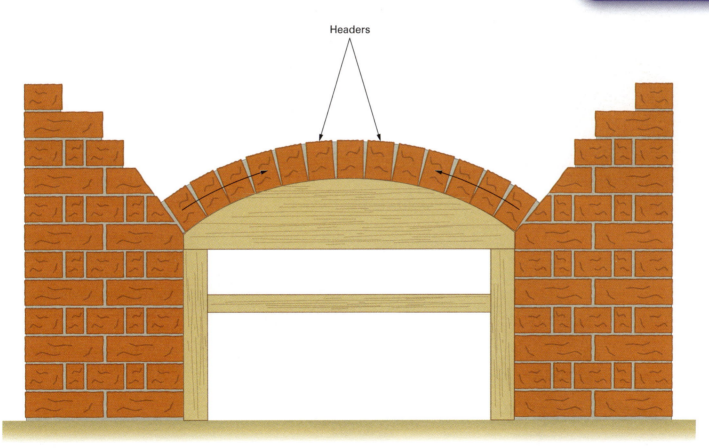

Figure 13.11 Headers used to form the arch ring

237

Constructing rough ringed arches

We will assume that the turning piece or arch centre has been correctly positioned and adequately supported with props and the folding wedges are in place.

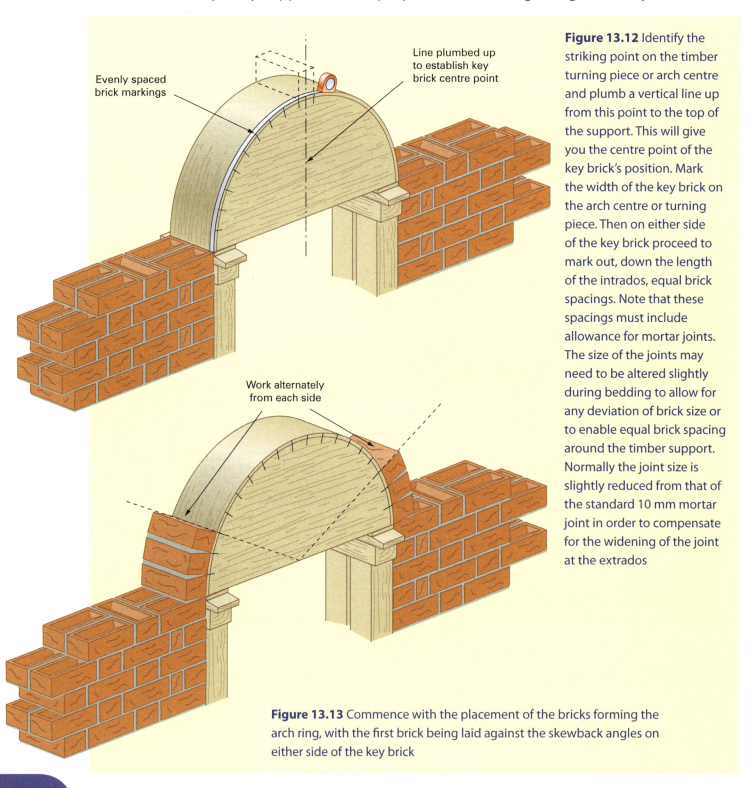

Figure 13.12 Identify the striking point on the timber turning piece or arch centre and plumb a vertical line up from this point to the top of the support. This will give you the centre point of the key brick's position. Mark the width of the key brick on the arch centre or turning piece. Then on either side of the key brick proceed to mark out, down the length of the intrados, equal brick spacings. Note that these spacings must include allowance for mortar joints. The size of the joints may need to be altered slightly during bedding to allow for any deviation of brick size or to enable equal brick spacing around the timber support. Normally the joint size is slightly reduced from that of the standard 10 mm mortar joint in order to compensate for the widening of the joint at the extrados

Figure 13.13 Commence with the placement of the bricks forming the arch ring, with the first brick being laid against the skewback angles on either side of the key brick

It is important to ensure that there is no bedding between the brick and the timber centre as this will result in difficulty in maintaining the correct curve of the arch bricks during construction and also stain the face of the bricks exposed once the support is removed. Packing may be introduced at the base of the joint being formed to allow for ease of pointing once the support is removed.

Bricks should be laid alternately on either side of the key brick to ensure that there is no overloading on any particular side of the support. Once the key brick position has been reached you must ensure that this brick is placed accurately in the marked position on the centre and that there are fully compacted joints either side of it.

Throughout the construction of the arch rings you must also maintain the face plane of the brickwork. This can be done by using a suitable, accurate straight edge or by erecting temporary line supports on each of the abutments. These can be built in brick and are known as **deadmen** (see Figure 13.14) or consist of timber or metal profiles, accurately gauged and plumbed.

Remember

Arch supports must not be removed until the bedding mortar has fully hardened

Brickwork NVQ and Technical Certificate Level 2

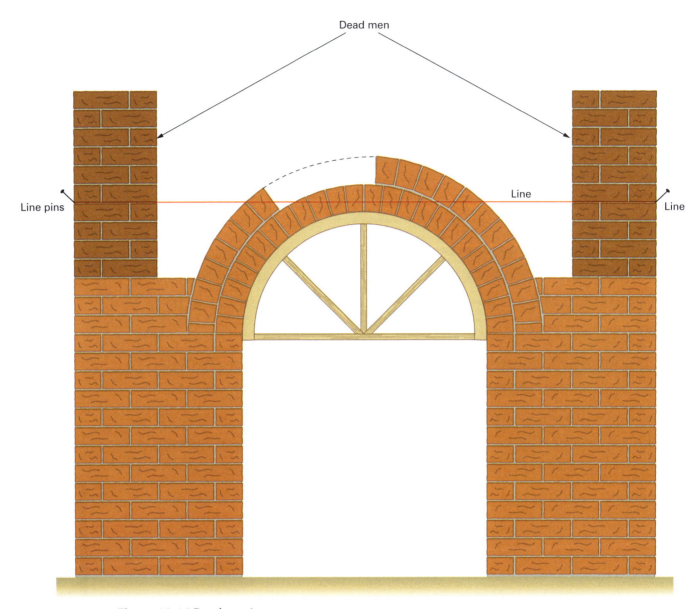

Figure 13.14 Deadmen in use

240

Chapter 13 Arch construction

Axed arch

This method is the total reverse to that of the rough ringed arch. In this method it is the bricks themselves that are cut to a wedge shape and the joints are uniform in shape and do not taper. These wedge shaped bricks are referred to as **voussoirs**.

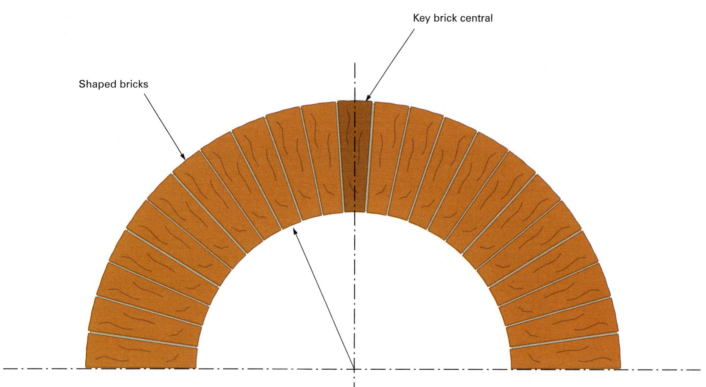

Figure 13.15 Section of an axed arch and the shaped bricks

On the job: Building a semi-circular archway to a garden wall

Max is setting out a semi-circular arch to form an archway to a garden wall. He has set the arch centre in place and is marking out the brick positions on the arch centre using a gauge of 75 mm. However, there is a 15 mm gap to place a brick into. How can he get over this to make the arch look correct?

Temporary support

Traditional methods of temporary support for arches during construction are:

- timber turning pieces
- timber arch centres.

Nowadays many sites use metal support systems or plastic arch formers, which are normally left in-situ (in place) following completion of the work. Other types of temporary support are also becoming more common and are gradually replacing the bulky, heavy traditional timber supports.

Where timber turning pieces or arch centres are used, these must be supported on both sides by props and placed directly on to folding wedges. Folding wedges enable small adjustments to be made to the support piece in terms of levelling. The wedges also allow for easy removal of the support piece once the mortar joints have fully set.

Chapter 13 Arch construction

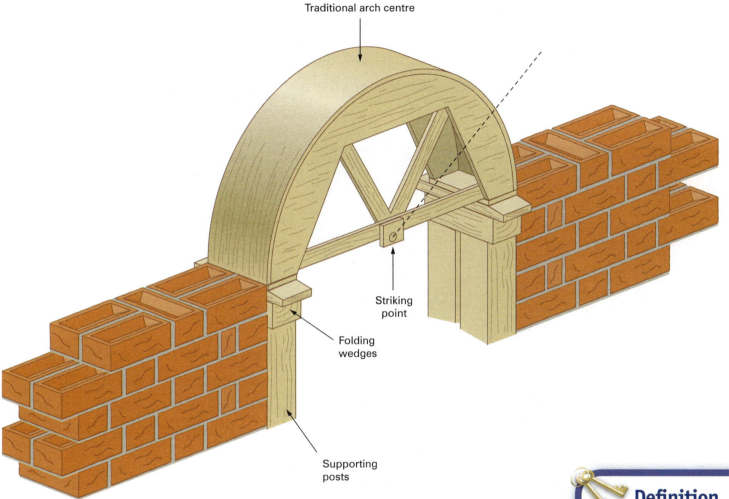

Figure 13.16 Traditional timber arch centre supported by props and folding wedges

On the job: Building a window opening with a segmental arch

Peter has built an opening to a cavity wall ready to receive a UPVC window with a segmental arch feature at the top. He has built the opening without using a **dummy frame** and is now putting the arch centre in place but finds it will not fit. What do you think has happened?

Definition

Dummy frame – a temporary frame put in place to ensure the opening dimensions and square are maintained while constructing brickwork

243

Complex arch types

It is possible to construct many different types of complex arches. In this section we will look briefly at a few arch varieties, although you will learn about them in more detail as you progress through higher levels of your qualification.

Gothic arch

There are many variations in the standard Gothic arch including:

- the lancet
- elliptical or tudor
- Venetian.

All Gothic arches have one thing in common and that is that they all have a pointed crown.

An example of Gothic arches

Camber arch

Camber arches are sometimes referred to as flat arches and are built with a slight skewback.

An example of a camber arch

Semi-elliptical (or three-centred) arch

A semi-elliptical arch is set out using three striking points, which give the appearance of the arch having three sections. A semi-elliptical arch can also be built as a five-centred arch.

An example of a semi-elliptical arch

FAQ

Why can't I remove the arch centre or support as soon as the work is completed?

The mortar needs time to set in order to make the arch strong enough to support the load above. If any support is removed too soon, the arch may move or even collapse.

Why are folding wedges used when positioning the arch support?

Folding wedges are used so that fine adjustments can be made to the height when positioning the arch support. They also make it easier to remove the arch centre when the mortar has set.

Knowledge check

1. Name three common arch forms.
2. Where would you find the intrados?
3. Why are folding wedges used?
4. What is the key brick?
5. What is the rise?
6. Which arch normally uses turning pieces?
7. Where is the striking point?
8. Draw a circle and then draw and label the following: radius; chord; segment; sector.
9. What are deadmen?
10. What have all Gothic arches got in common?

chapter 14
Pointing and jointing

OVERVIEW

This chapter is based on the finishing of brickwork after it has been built, as well as rejointing to existing work, whether on a new building if the joints have been raked out, or on an older building where the joints have badly weathered and broken down. As with all face brickwork and blockwork, once completed the joints are finished to weatherproof the wall and give a pleasing appearance. This chapter shows how this is carried out and the different ways walls can be finished.

This chapter will cover the following:

- Health and safety
- Jointing
- Pointing
- Raking out
- Mixes
- Repointing
- Types of joint finish.

Health and safety

The main health and safety issues with pointing and jointing are skin contact with the mortar and the risk of mortar splashing into the eyes when using it. With repointing, the raking out process involves the use of grinders and other equipment which create dust. Appropriate PPE (personal protective equipment) such as gloves, safety goggles and a respiratory mask should be worn.

Jointing

Jointing is carried out to make sure the mortar joint is completely filled and in contact with each brick. This will prevent rainwater penetration and possible damage from frost, which would crumble the joint. Another reason for jointing is to give the overall appearance of the wall a pleasing look to the eye. Once the jointing is completed, rainwater should run down the surface of the wall and not into any gaps left in joints.

On larger sites, before commencement of any brickwork, sample brick panels may be constructed and pointed to show how the finished product will look, giving architectural staff a chance to change the type or colour of pointing required. This is because sometimes coloured mortar may be used for the joints.

One of the most important things to consider is at what time jointing should be carried out. As the brickwork is carried out over the course of the day, lower courses will dry out but other factors need to be taken into consideration:

1. The type of brick used – stock or engineering bricks do not absorb moisture as fast as some softer bricks. Therefore the joints will take longer to dry to the required texture for jointing.

2. The moisture content – the bricks may be damp before use. Therefore moisture absorption will be slow.

Find out

Who would normally request that sample brick panels be built?

Chapter 14 Pointing and jointing

3. The weather – when working in summer the heat will dry out joints faster, due to the warmth of the bricks, than in damper conditions, when moisture stays in the air.

Joints need to be checked by touching to see if they are ready for jointing. If they are too dry it will be difficult to carry out jointing and if too wet the joint mortar is inclined to 'drag', giving poor adhesion to the bricks and a poor quality appearance.

This operation has no given time span, so understanding and experience are key to knowing when is the right time.

Pointing

Pointing is the process of joint filling to brickwork when the mortar joints have been previously raked out to approximately 12 mm (whether to newly built work or to older brickwork joints which have weathered over the years). This type of work requires more skill and patience than jointing as it is very time-consuming and great care must be taken not to smear the new mortar onto the face of the brick.

There are different types of joint finishes, which we will cover in this chapter. Some take longer to carry out than others, hence the cost of pointing varies considerably as some types can take twice as long to complete.

Remember when pointing:

1. Always start from the top and work downwards – dropped mortar, or mortar brushed off, can fall on to already completed work.

2. Make sure the joints are clean of any loose old mortar.

3. Brush the area to be pointed to remove any dust.

4. Wet the wall so that the bricks absorb the water to give good adhesion for the new mortar. Some bricks will require more water than others.

> **Did you know?**
>
> Some bricks can 'swim' with water and may require pointing the next day. To 'swim' means to float on the bed joint due to delayed drying out time

249

5. Apply the mortar filling to the **perps** first so as to keep a continuous bed joint when applied.

6. When sufficiently dry, brush off with a fine brush to remove any excess particles of mortar.

Raking out

Joints need to be raked out to a depth of between 15 and 18 mm. This can be carried out in several ways:

1. By using a racking out tool (or chariot) set to the required depth to rake out the joints evenly. This is perfect for very soft joints but may not be for harder or variable joints.

2. By using an angle grinder, but great care and expertise is required not to touch and mark the brick faces or damage the arrises of the brick as joint depths can be varied. Dust from the cutting can cause problems if working in a built-up area (dust gets on to people's property).

3. By using a bolster, chisels or comb chisel, but again great care should be taken not to damage the arrises of the brick and the depth could be varied.

Sometimes a combination of the above may be the best solution. Once raked out, pointing can be carried out.

Mixes

When mixing mortar for pointing, fine sand should be used to give a smoother finish, especially if the joint is to be cut with a trowel. This is because any edges would break unevenly if coarser sand were used. If pointing is being carried out to relatively new brickwork, it would be better to use the same mix as for the original works. Sometimes pointing may be done, allowed to dry, raked out and a different coloured mortar used.

This is usually done because of the time aspect (i.e. the brickwork could have to be finished, then raked out for future pointing so as to allow the other trades to progress).

Different types of bricks will require different mixes. Engineering bricks will require a strong cement-based mix whereas soft clay bricks will require a weaker mix, possibly with lime incorporated. Table 14.1 shows a basic breakdown:

Group 1	Class A Engineering Bricks	1:3 or 1:4
Group 2	Class B Engineering Bricks	$1:\frac{1}{2}:4\frac{1}{2}$
Group 3	Face Bricks	1:5 or1:1:5 or 1:1:6

Table 14.1 Mortar mixing ratios for different types of brick

Lime is more likely to be used on older buildings where the mortar originally used was sand and lime but with cement added to give extra strength and durability. In most cases the mix should not be stronger than the original material used.

The mixes shown in Table 14.1 can also be used for jointing mortars.

Repointing

Repointing is the process carried out on older properties where the mortar joints have broken down due to rain, wind or frost exposure over many years. This could be because the original mortar was very soft with no cement being used.

Another reason for repointing could be because of insect attack. Soft joints allow small insects to burrow into the joints, the worst being from Osmia Rufa or the Mason Bee, which burrows into soft joints to lay its eggs. It has been known for this type of bee to literally demolish a building by removing so much mortar as to make it dangerous to enter. Walls can be spray-treated to deter this type of attack and in most cases, if repointed with cement-based mortar, attacks cease as the new joints are not as soft.

Brickwork NVQ and Technical Certificate Level 2

Did you know?

Mason Bees do not like cement-based joints

Mason Bee in soft joint

On the job: Repointing on an older property

Amir has been asked to cut out and repoint a section of wall on an old property that has been patched by the previous owner using nearly neat cement. What tools should Amir use to cut the pointing out and what problems could arise?

Types of joint finish

Ironed or tooled joint

This type of joint is the most commonly used as it covers up slight impurities in the brick arrises and is the quickest to carry out. See Figure 14.1. There are different sizes so care should be taken to use the same size each time. Smaller sizes give a deeper profile whereas the larger diameters give a shallower, rounded look. The jointing is carried out as work progresses.

Figure 14.1 Tooled joint

Recessed joint

With this type of joint the mortar is dragged out to a minimum depth of approximately 6 mm by using the chariot or a similar recess tool. Great care should be taken to ensure all of the joint is removed. A recessed joint is better used when the bricks are a harder, more frost-resistant type as water can lie on the edge of the recessed arris on the brick. Care must be taken to ensure all the joints are full before commencing the raking out process. Again, jointing is carried out as the work progresses.

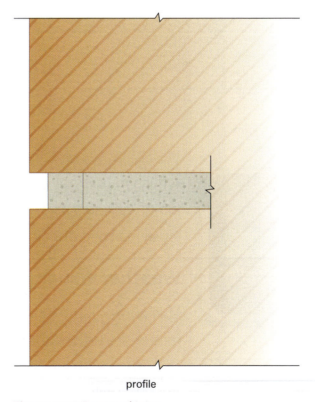

Figure 14.2 Recessed joint

Flush joint

This type of joint gives a simple look but it is quite difficult to keep a complete flush surface finish (see Figure 14.3). If modern type finishes are not required, it gives a **rustic** look which may be more in keeping with the surroundings. This type of joint is carried out by using a hardwood timber or plastic block to smooth and compact the surface of the mortar into place.

> **Definition**
>
> **Rustic** – old and natural looking; traditional

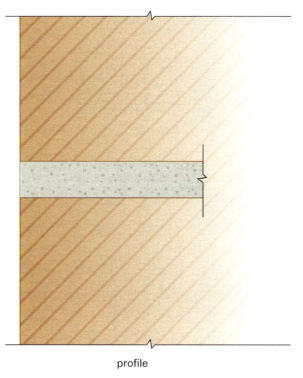

Figure 14.3 Flush joint

Figure 14.4 Struck joint

> **Note**
>
> If on a large site, make sure all bricklayers 'strike' to the same side on the perp joints

A flush joint is not ideal if the bricks used are not regular in shape as the joints will show any deviation and could look wider than they are. Jointing is always carried out as work progresses.

Weather struck joint

This type of joint is slightly sloping to allow rainwater to run down the face of the brick rather than lie at the joint (see Figure 14.4). The mortar is smoothed with a trowel, with the mortar the thickness of the trowel below the top brick and flush with the brick below. The same process is carried out with the perps and the left side of the joint is below the surface.

Weather struck and cut pointing

This is the most common type of jointing carried out on previously raked out joints (see Figure 14.5). It can cover any irregularities to bricks, creating a straight appearance. This type of pointing is the hardest and most time-consuming to do. The mortar is smoothed flush to the brick at the top of the joint and about a trowel thickness proud of the brick at the bottom. The mortar is then allowed to dry slightly and is then 'cut' in a straight line using a tool called a Frenchman.

The straight edge should be kept off the wall using cork pads, nails or screws, so that the cut excess can drop and not be squashed against the face of the wall. Perps are again angled to the left and finished in the same way as the bed joints, but cut with a pointing trowel. All perps should be completed before the bed joint so as not to mark the beds with the trowel and to keep a continuous joint to the bed. This should then be lightly brushed sideways so as not to drag the edges of the bed joints.

Reverse struck joint

This type of finish is normally used for internal walls, giving a smooth finish to work that is not plastered as no shadows appear at the joint area. Care must be taken to ensure the bottom edge is flush to the brick to stop the joint becoming a dust trap when work is complete. The wall would then be given a paint finish.

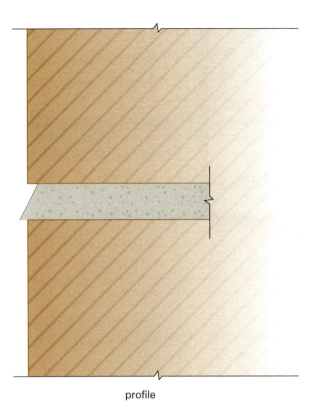

Figure 14.5 Weather struck and cut joint

Figure 14.6 Reverse struck joint

Brickwork NVQ and Technical Certificate Level 2

Did you know?

Sometimes a glue (PVA) can be added to the mortar to give a better adhesion to the brick and help bond to the existing mortar

With all types of pointing, it is essential that the new mortar bonds well to the existing material, otherwise moisture will be trapped between and, with frost, will break the newer material down, resulting in more costly repairs.

FAQ

What is a half-round profile joint (also known as a bucket handle finish)? Why is this type of joint, along with a tooled joint, the most commonly used method when pointing?

A half-round profile is a shallow, round, inwards joint finish so-called because bricklayers used to use a bucket handle to create it. A special tool can now be bought to do this instead. The reason this type of joint along with a tooled joint are the most commonly used is simply because they are the quickest and therefore the cheapest to do.

Knowledge check

1. Describe jointing.
2. What mix should be used on standard face bricks?
3. What is the most commonly used type of jointing?
4. Describe pointing.
5. Name an insect that burrows into soft joints
6. What type of mortar was often used on older properties?
7. What is the most commonly used type of pointing to previously raked out work?
8. What is the minimum depth that mortar joints should be raked out on new work for later pointing?
9. On what type of jointing would you use a chariot on fresh mortar?
10. What tool is used to cut mortar on weather struck and cut pointing?

chapter 15

Drainage systems

OVERVIEW

Drainage systems are pipe work, usually underground, which carry away waste matter and water from a building. They are a very important part of the construction process, as all waste materials have to be carried from a building safely to prevent the spread of disease.

There are two main types of waste carried in drainage systems: foul waste and surface waste. Foul waste consists of waste from toilets, baths, sinks etc. whereas surface waste is rainwater. If foul waste were not collected correctly raw sewage would be running in the streets, as it did many years ago. With populations expanding, many larger cities have to renew drainage systems to cope, as systems put in 200 years ago, as in London, were never meant to carry the capacity.

This chapter breaks down these systems to show how they are installed, and the methods used.

In this chapter we will look at:

- Types of drainage system
- Drainage ventilation
- Drainage pipes and fittings
- Gradients of drainage
- Pipe bedding and surrounds
- Safety in excavations
- Inspection chambers
- Pipe cutting
- Pipes entering buildings
- Drain testing
- Suggested procedures for constructing a drainage system.

Brickwork NVQ and Technical Certificate Level 2

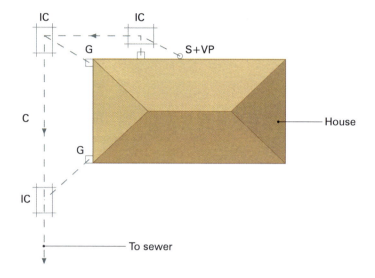

Figure 15.1 Plan of combined drainage system

C = 100 mm diameter combined drain
S+VP = Soil and vent pipe
G = Gully
R.W.P. = Rain water pipe
IC = Inspection chamber

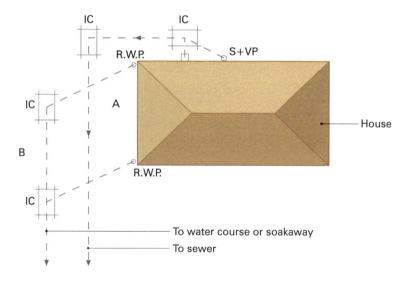

A = 100 mm diameter foul drain
B = 100 mm diameter surface water drain
C = 100 mm diameter combined drain
S+VP = Soil and vent pipe
G = Gully
R.W.P. = Rain water pipe
IC = Inspection chamber

Figure 15.2 Plan of separate system

It is required that a drain should be:

- designed to suit its intended purpose
- laid out as simply as possible
- sloped enough to take away waste
- airtight and/or watertight
- able to prevent gases entering the building
- ventilated to get rid of the gases
- accessible for inspection and cleansing.

Types of drainage system

There are two types of drainage system:

- the combined system
- the separate system.

The combined system

In this type of drainage system the foul and surface water is conveyed in the same pipe.

The separate system

In a separate system the foul and surface water are kept separate by the use of two drainage systems. The foul water is discharged into a company sewer (local water authority) or type of container, and the surface water is carried away to a water course (canal, river, stream etc.) or a soakaway.

Chapter 15 Drainage systems

In the modern era great planning is undertaken to collect and **treat** as much waste water as possible to meet the demands of every one of us. Whether it is for drinking, bathing, or keeping our clothes clean, a vast supply of fresh water is required in every country just to survive. All of this is generated through climatic change (rainfall etc.).

Drainage ventilation

Drainage ventilation is very important for two main reasons:

1. to allow gases to escape from the drain
2. to equalise pressure so that the air pressure in the drain is the same as outside the drain.

In a building with an upstairs toilet etc. the outlet connection joins on to the soil and vent pipe, which then carries on above the roof level, to vent the gases. A pipe guard is put in to the top to prevent birds nesting at the top from causing a blockage.

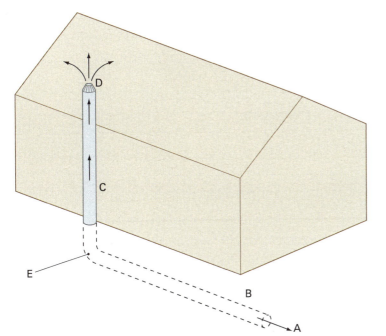

A = To sewer
B = 100 mm diameter drain
C = Soil and vent pipe
D = Pipe guard
E = Rest bend

Figure 15.3 Drainage ventilation

Definition

Treat – the treatment process to change waste water into clean drinkable water through several stages of purification

Did you know?

Some countries with hot climates do not have the required rainfall or facilities to collect and recycle wastewater. Drought is therefore a major concern, as well as disease caused by poor sanitation. These problems are usually due to a lack of money and resources to provide the necessary systems required

Did you know?

If drains were not ventilated, inspection covers could blow under the pressure

259

Drainage pipes and fittings

Pipes

The materials used for pipes are:

- clay
- UPVC
- concrete
- iron.

The minimum diameter of a pipe carrying soil water is 100 mm.

Most straight pipes are circular in section and contain a slight chamfer on each end to enable easy entry into the joining coupling.

Did you know?

Most drainage systems (such as those for domestic properties) adopt 100 mm minimum size pipes. Other sizes can be specified dependent on the use

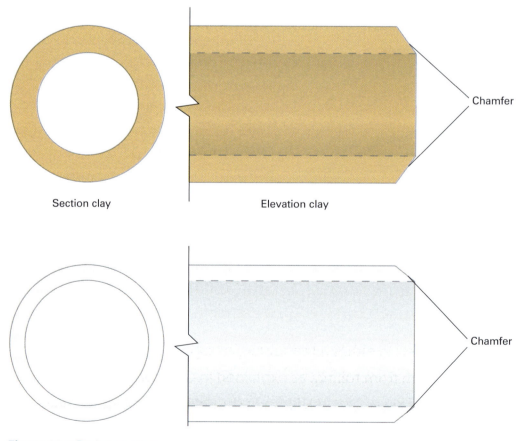

Figure 15.4 Drainage pipes

Chapter 15 Drainage systems

Fittings

These are components used in drainage other than straight pipes. There are numerous fittings available – some common ones are shown in Figure 15.5.

Fittings are usually the same material as the pipes being used but can differ with various adapters.

The main locations for fittings are:

- at a junction
- at a bend
- at the end of a drain
- in an inspection chamber.

- Used when a branch drain meets a main drain
- Not recommended on soil drainage

- Used when a drain changes direction
- Should be avoided where possible on soil drainage

- Used to terminate the end of a drain and receive sink wastes and remainder pipes
- Water seal remains in bottom of gully to prevent gases excaping

- Used in an inspection chamber where a branch drain meets a main drain
- Aids cleansing and inspection

Figure 15.5 Types of drainage fitting

Remember

Before deciding on any fittings required for a job, always obtain a manufacturer's catalogue

Storage of pipes and fittings

The storage of pipes and fittings usually depends on what material they are made from.

Clay

Pipes should be stored in either a triangular pile or in specially shaped racks to prevent them from rolling, as explained in Chapter 6 Handling and storage of materials (see page 135). Fittings should be stored individually and not on top of one another, preferably on shelving in a secure place.

Remember

The quality of materials in drainage has to be of the highest standard so careful handling and storage is essential to avoid damage to the pipes and fittings

261

UVPC

Pipes should be stored as for clay. UVPC is not as prone to damage as clay but should be protected from bright sunlight and frost. Fittings are light and robust but should be stored with care and kept as clean as possible.

Concrete

The storage of concrete pipes is the same as for clay pipes.

Iron

Iron pipes have a protective coating so there is no need to protect them from moisture.

Rigid pipe joints

Rigid joints refer to joints made with sand and cement (in the ratio of 1:2). The spigot is inserted into the socket and 'centred' by tarred yarn or rope. This joint has the disadvantage that the slightest movement of the pipe could fracture the joint and cause a leak.

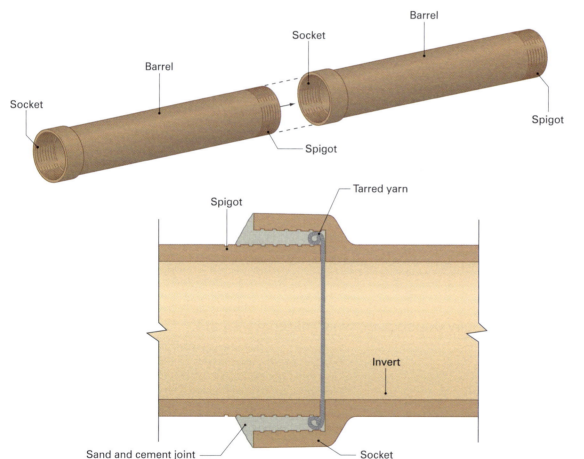

Figure 15.6 Rigid pipe joints

Flexible pipe joints

Flexible joints have many advantages over rigid joints as they:

- can move slightly without leaking
- can be laid faster and more efficiently
- are not damaged by freezing temperatures
- can be tested immediately after laying
- are easily aligned.

Flexible drainage pipe joint

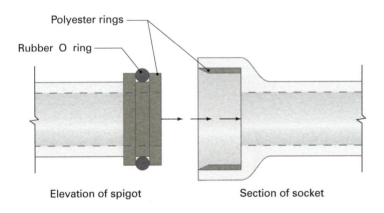

Hepseal flexible joint - In this joint the O ring is lubricated and the spigot is pushed into the socket compressing the O ring.

Hepsleve flexible joint - In this joint the spigot of the pipe is lubricated and pushed up to the centre stop of the coupler, thus compressing the rubber sealing ring.

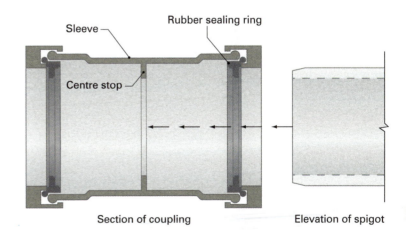

Figure 15.7 Flexible jointing

Gradients of drainage

For a drain to work efficiently, the right slope is essential. The gradient is a ratio and is obtained by dividing the length of the drain by the amount of fall. For example, if a drain run is 20 m long and falls 1m then the gradient is 20 ÷ 1 = 1:20.

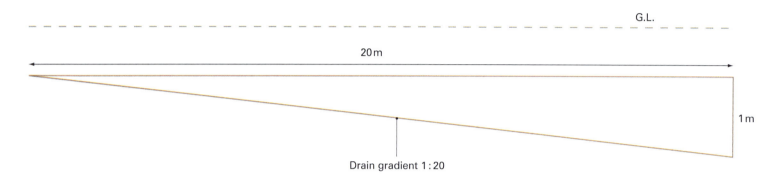

Figure 15.8 Drain gradient

Remember

For all sizes of pipes there is a gradient that is most efficient. For a 100 mm diameter pipe the best gradient is 1:40, for a 150 mm diameter pipe 1:60 and for a 225 mm diameter pipe 1:90

To maintain the correct gradient of a run of pipes, site rails are set up at the end or at intervals along the drain and the invert levels of the drain. Pipes can be checked by the means of a traveller (a T-shaped rod).

A traveller

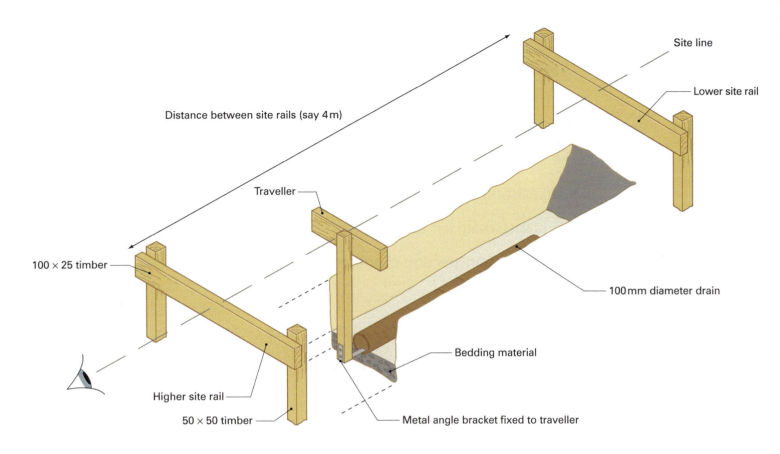

Figure 15.9 Gradient of drain showing traveller

Maintaining gradients

In order to maintain gradient:

1. Fix lower site rail (any height).
2. Divide distance between site rails by gradient (1:40) = 4000 ÷ 40 = 100 mm.
3. Set higher site rail 100 mm higher than lower site rail.
4. Determine length of traveller at lower end of drain (invert level to site line).
5. Lay each pipe and sight traveller as shown to maintain gradient.

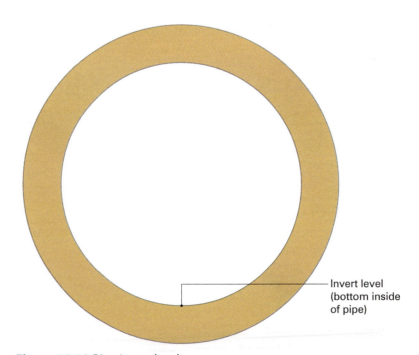

Figure 15.10 Pipe invert level

Setting out the gradient of a drainage system

To do this operation a certain amount of information is required before you start. It is acceptable to start at the lowest point and work to the highest so the starting point must be known.

The starting point must be:

- an existing local authority sewer
- a branch drain left on the sewer
- an existing inspection chamber
- a septic tank.

When the starting and finishing points have been found the 'path' of the drain can be established.

Invert levels

When calculating gradients the invert of the pipe is used as the reference point and this can be related to the site datum. When a drainage layout is

Figure 15.11 Invert levels at inspection chambers

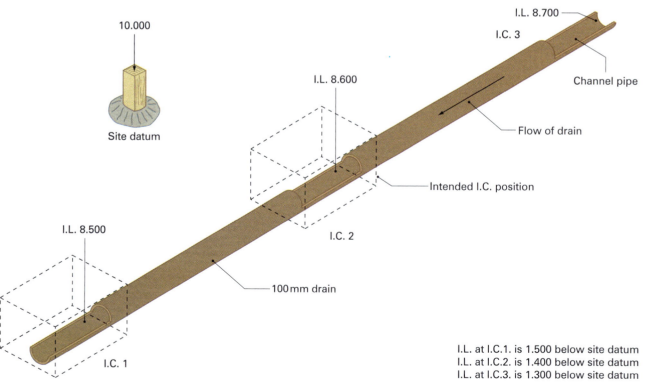

I.L. at I.C.1. is 1.500 below site datum
I.L. at I.C.2. is 1.400 below site datum
I.L. at I.C.3. is 1.300 below site datum

designed, the architect or draughtsman will usually give the invert level at each manhole, which will of course contain channel pipes. The invert level will be in metres to three decimal places e.g. 7.675 m. The inspection chambers on a drain are placed at changes of direction or gradients so it is practical to place site rails near inspection chambers.

Gradient boards

Another way of setting out a gradient is by the use of a gradient board. It is a tapered piece of timber, which is placed onto pegs within the trench or laid on the pipes after laying to check the gradient. A spirit level is used on top of the gradient board to check the position.

Remember

Gradient boards are only really suitable for short drain runs and would not be accurate on long runs

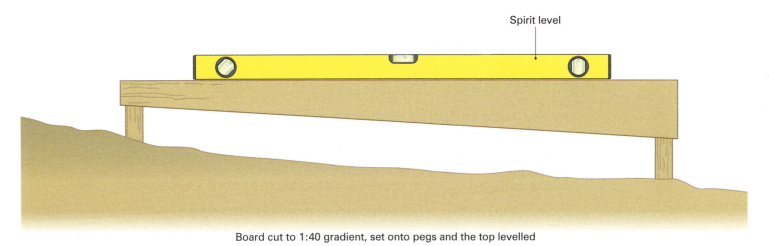

Board cut to 1:40 gradient, set onto pegs and the top levelled

Figure 15.12 Gradient board on pegs

Pipe bedding and surrounds

Pipe bedding and surrounds are very important as they have a two-fold purpose. They must:

1. protect the pipe from receiving any loads that may break or damage the drain
2. allow movement of the pipe within the ground.

Did you know?

Flexible pipes are normally bedded and surrounded with pea gravel but in some cases solid concrete is used

Safety in excavations

Safety is of prime importance before and during drainage work so always ensure you are wearing the correct PPE (personal protective equipment).

Considerations before excavating

Existing underground services and obstructions that might be in the way must be considered, for example:

- water mains
- gas mains
- electricity cables
- telephone cables
- TV cables
- tree roots and other buried objects
- drains.

Some existing services can be located by means of an electronic cable locator and as much information as possible should be gathered to avoid damage and expensive repairs.

Other hazards

Some grounds contain gases (methane) and there must be an awareness of this before starting work so that necessary precautions can be taken. Another important consideration in excavations is the lack of oxygen. This can occur in confined spaces or in deep excavations, and sometimes tests have to be carried out to monitor the oxygen levels.

Safety around the trench

There are some very simple things you can do to make sure that the trench is as safe as possible:

Chapter 15 Drainage systems

- Keep excavated material cleared away from edges of trenches to stop the excess weight causing the ground to collapse.
- Store pipes at right angles to the trench so they do not roll in.
- Put up warning signs and guardrails as required to protect workers.

Trench timbering

Trench timbering is the operation of supporting the sides of a trench. In drainage, operatives will spend a considerable amount of time in the trench during the laying of the pipes, so their safety must be ensured by the prevention of trench collapse. Drain trenches can be quite deep and all excavations 1.2 m deep and over should have some kind of support. Even shallow trenches may need support in certain soils, and remember that operatives may have to kneel or lie down to carry out some operations.

Did you know?

Most operations now use steel sheets with timber bracing as this is much stronger than timber alone

Trench timbering

On the job: Laying drainage pipes

Wendy is about to lay a new drainage run from the first inspection chamber and connect it to the main road sewer. The sewer pipe is 3 m deep and the trench has been excavated ready for the pipes to be laid. There was heavy rain the previous night and the trench sides have not been supported. What should she do?

Inspection chambers

Inspection chambers (ICs) are compartments built over a drain at certain points to provide access for inspection and cleansing.

What materials are used for ICs?

- brickwork
- plastic
- pre-cast concrete
- **in-situ concrete**.

Where are ICs positioned?

- at changes of direction
- at junctions
- at the end of a drain
- on very long straight runs
- just before crossing a boundary.

Definition

In-situ concrete – concrete placed in its finished position

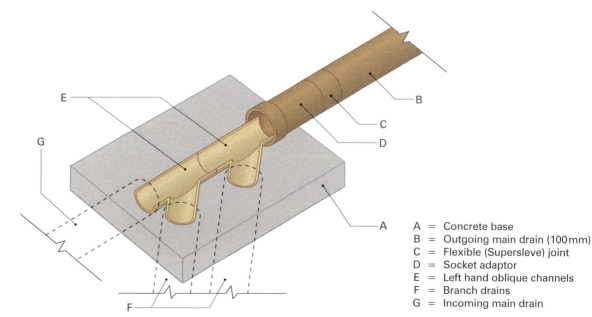

A = Concrete base
B = Outgoing main drain (100 mm)
C = Flexible (Supersleve) joint
D = Socket adaptor
E = Left hand oblique channels
F = Branch drains
G = Incoming main drain

Figure 15.13 Base of inspection chamber

Starting off an inspection chamber

Mortar for brickwork should be made with fresh Ordinary Portland Cement or, in certain grounds, sulphate-resisting cement could be specified.

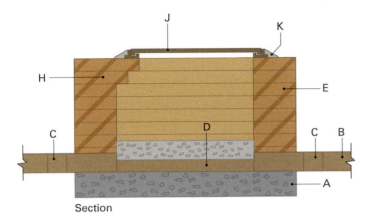
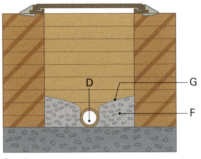
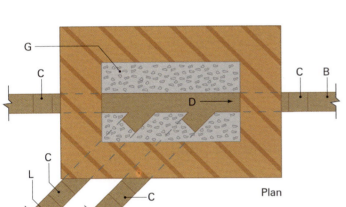

A = Concrete base
B = Main drain
C = Flexible joint
D = Channel pipes
E = One brick walling in class A or engineering bricks
F = Concrete benching (should rise vertical to crown level of incoming pipe)
G = 1:3 cement and sand finish to benching sloping to 1:12 with 25 mm radius
H = Corbels to reduce I.C. to cover size
J = I.C. cover and frame
K = Mortar fillets
L = Branch drains

Figure 15.14 Inspection chamber details

Constructing a brick IC

The following method describes how a brick IC would be constructed using clay pipe drainage:

1. The drainage system will be laid first with channel pipes at IC position.

2. Lay 150 mm thick concrete base to IC walls.

3. Construct brick walls (minimum one brick thick) to desired height, observing the following:

    (a) all channels to be in the IC walls

    (b) minimum size 600 x 450 mm internal

    (c) joint IC in weather struck or half-round finish

    (d) build in step irons as work proceeds at 300 mm centres vertically and horizontally

    (e) if IC larger than 600 x 450 mm, corbel brickwork near top to receive IC cover and frame

    (f) brickwork should be in English or Water bond.

4. Form benching in concrete and finish as specified.

Always make sure an IC is cleaned out on completion and no debris is left in the channel or adjoining junctions.

Junctions

There are three main ways of joining a branch drain into a **main drain**:

1. by a channel junction

2. by a half or three-quarter channel (very often used to join into existing ICs)

3. by a saddle: a special fitting that fits in a hole in the main drain or sewer and is made good with mortar.

> **Definition**
>
> **Main drain** – a local authority main sewer

Access to inspection chambers

Galvanised step irons are built in as the work proceeds in ICs 1 m deep or more.

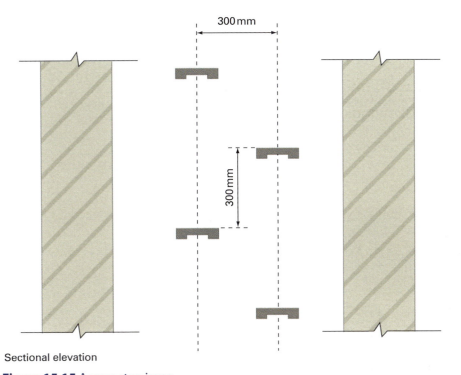

Sectional elevation

Figure 15.15 Access step irons

Step irons are built in as brickwork progresses

Pipe cutting

Clay pipes

When pipes require cutting they should be cut with a chain cutter as shown in the photo. The pipe is marked off at the required length. The chain is wrapped around the pipe and then tightened by either a screw thread or scissor action depending on the type used. This shears the pipe by compression, giving a neat square cut.

The pipe should then be chamfered by use of a pipe trimmer or a disc cutter.

Pipe being cut with a chain cutter

If static or portable disc cutters are used, all relevant regulations and safe practices must be followed, such as:

- correct disc fitted by a **competent person**
- instruction on how to use the cutter properly
- any necessary guards are in place
- grade 1 impact eye protection for the operation
- warn and safeguard nearby operatives
- be aware of any necessary noise controls.

Definition

Competent person – in these circumstances, usually a person holding an Abrasive Wheels Certificate

Plastic pipes

Plastic pipes can simply be cut with a fine tooth saw. The chamfer can be re-formed by filing or planing and any burrs can be removed before assembly.

Pipes entering buildings

Where a pipe passes through a wall to enter a building, certain precautions are necessary to prevent damage to the drain:

- The wall above the pipe must be adequately supported to prevent any weight of the structure settling on the pipe.
- A 50 mm gap should be maintained all around the pipe.
- A rigid sheet material should be placed both sides of the hole to prevent vermin entering the building and backfill entering the 50 mm clearance.

Did you know?

Concrete lintels are normally used as support when bridging an opening

> **Remember**
>
> Air and water tightness tests should be carried out before backfilling takes place, as obviously any necessary repair of leaks would have to be undertaken

Drain testing

There are several ways of testing drains for air or water tightness. It is an extremely important operation as foul water leaking out of a drain can cause a serious health hazard. A statutory inspection by the local authority must be performed and the guidelines of the Building Regulations must be followed.

Water (hydraulic) test

When the water is test is carried out, all parts of the system are filled with water. Air must be allowed to escape and then, two hours after filling, the level of the water is topped up. Leakage over the next half an hour should be minimal. Allowance is made for evaporation when carrying out the test in hot weather, as pipes will be in direct sunlight for the duration of the test.

Once the drain has been checked and inspected, the run can then be backfilled, ensuring a minimum covering of pea gravel protection to the pipes before any soil. Care should be taken in backfilling to ensure that the pipes are not damaged by stones etc. during this procedure, and further testing should take place after this process.

Air test

Where water is unavailable or the drainage system is too large to be filled with water, an air test can be used. This is done by sealing off the ends of the drain and pumping air into the drain to pressurise it. The pressure is read on a U tube (a Manometer) and any severe drop in pressure would suggest a leak.

Suggested procedures for constructing a drainage system

Whichever type of drainage system you are laying and whichever type of material the pipes and fittings are made of, below are the basic procedures you should always follow:

1. Use the manufacturers' catalogues for ordering materials.

2. Order accurately as fittings are very often unusable elsewhere.

3. Handle and store pipes and fittings carefully to prevent damage and expensive waste.

4. Set out pipe runs and erect necessary site rails.

5. Construct traveller.

6. Be aware of any existing underground hazards.

7. Excavate trenches and IC position.

8. Provide trench support if required.

9. Lay pipes and channels at IC position.

10. Lay pipes on specified bedding materials.

11. Construct ICs.

12. Test drains.

13. Surround and backfill to drain pipes.

14. Re-test drains after backfill.

FAQ

Why are drain systems tested twice?

The first test checks for leaks before backfilling (filling the excavation) and the second test makes sure no pipe work has been damaged during backfilling. Clearly you don't want to find any leaks after backfilling so filling the excavation must be done with care.

Knowledge check

1. What does IC stand for?
2. What is the best method to cut plastic pipes?
3. What should be used as bedding material for pipes?
4. What gradient is used for 100 mm pipes?
5. What is a rigid joint?
6. Why should excavated material be kept away from trench edges?
7. What are the two types of drainage system?
8. At what depth should trench timbering be used?
9. What is the minimum thickness of wall for an IC?
10. Why are drainage systems ventilated?
11. Why should the ground be checked before any excavation?
12. What is the invert level?
13. What are the two main types of drain testing?
14. What is a gradient board used for?
15. Name two types of materials used for drainage pipes.

chapter 16
Basic concreting

OVERVIEW

Concrete is one of the most important man-made products in the construction industry, forming the very foundation for all weight-bearing structures within the industry. Most people use it every day in different ways, whether they are walking on a footpath, driving on motorways, crossing a bridge, or working in an office. If no concrete foundations were in place, these tasks would be a lot more difficult to do. People take for granted the importance of concrete and the different areas in which it is used, as in most cases the finished item is never seen. As a bricklayer you will need to understand how it is made, the different strengths required, the different finishes that can be made and, most of all, the main areas in which it is used.

This chapter will cover the following:

- Health and safety
- What is concrete?
- Mixes
- Types of concrete
- Formwork
- Reinforcement
- Expansion joints
- Workability
- Compacting
- Surface finishes
- Curing
- The Cement & Concrete Association.

Health and safety

The main health and safety legislation covering concrete and its use are:

- The Health and Safety at Work Act
- Control of Substances Hazardous to Health (COSHH) Regulations.

The Health and Safety at Work Act

Under the Health and Safety at Work Act, it is the duty of the employer to provide you with suitable working environments, supervision, information and training that will keep you safe while you are working. In addition, it is the duty of every employee to work in a safe manner and not put other employees at risk as a result of their actions. The Health and Safety Executive (HSE) enforces the act to help prevent accidents occurring in the workplace.

COSHH

The COSHH Regulations cover the safe use and storage of all materials and are also enforced by the HSE.

For more information on both the Health and Safety at Work Act and COSHH Regulations, look back at Chapter 2 Health and safety, page 33.

> **Remember**
>
> The HSE is there to give advice and help as well as enforce laws and regulations

What is concrete?

Concrete is a mixture of cement, aggregates and water. The aggregate is normally in two parts:

1. fine aggregate – sand and limestone dust
2. coarse aggregate – gravel or limestone chippings.

The coarse aggregate is the bulk of the concrete, while the fine aggregate fills in the voids between the larger particles.

> **Remember**
>
> All aggregates should be 'well graded', meaning that they range from small to large grains so they fill in all the voids in the concrete

The cement is the binder that holds all the aggregates together. Water is required to cause a chemical reaction (**hydration**) that changes the dry cement powder into an adhesive.

Cement

The most common type of cement used for concrete is Ordinary Portland Cement (OPC). Over the years, however, several types of Portland cement have been developed which can be used for concrete in more specified construction. These include:

- Rapid-Hardening Portland Cement (RHPC), which gives strength more quickly than OPC

- Sulphate-Resisting Portland Cement (SRPC), which should be used in ground containing a high level of sulphates (this is because sulphates damage OPC)

- Low Heat Portland Cement (LHPC), which produces less heat during hydration. This makes it more suitable than OPC where large masses are needed that could produce high temperatures, which might lead to the concrete cracking.

Fine aggregates

Fine aggregates can be obtained from riverbeds, sand pits or dredged from the seabed. Dredged aggregates must be washed to remove any mud and weed. Seashore sand must not be used due to the high salt content.

The shape and size of the grains can affect the strength of the concrete. The particles should be irregular in shape (not rounded), well graded with the size not larger than 5 mm. Also, if the grain size is too small it will increase the total surface area so that the designed quantity of cement becomes insufficient, causing a weakness in the finished mix.

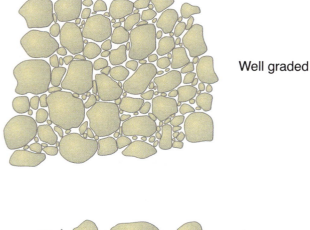
Well graded

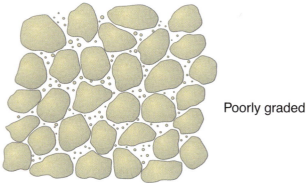
Poorly graded

Figure 16.1 Well graded and poorly graded concrete

Did you know?

Masonry Cement must never be used for concrete as this contains a plasticiser, which results in a weak concrete mix

Remember

Always use the correct type of cement for the job

Clay and silt particles also prevent the cement bonding to the aggregate, so each load delivered to the site should be tested for 'cleanliness'. The amount of silt must not be more than 10% of the volume of aggregate. The silt test is used to check this.

Silt test

Materials and equipment

- Sample of sand
- Water
- Salt
- Glass jar or measuring cylinder
- Tape measure.

Method

Step 1 Place 25 mm of water into the jar, add 1 teaspoon of salt and gradually add the sand until the level of the top of the sand reaches 50 mm.

Step 1 Ingredients being added to water

Step 2 Shake the jar for one minute.

Step 2 Jar after shaking

Step 3 Leave to settle for three hours. Measure the height of the aggregate and the thickness of the silt layer.

Step 3 Jar after three hours

To work out the percentage of silt in the aggregate the following sum must be calculated:

$$\frac{\text{Thickness of silt}}{\text{Total height of aggregate and silt}} \times \frac{100}{1}$$

For example, if we measure the sand to be 45 mm and the silt thickness was 5 mm:

$$\frac{5}{45 + 5} \times \frac{100}{1}$$

$$\frac{5}{50} \times \frac{100}{1}$$

$$0.1 \times 100$$

$$= 10\%$$

Conclusion – this sample would be suitable.

> **Remember**
>
> As with fine aggregate, the coarse aggregate should be well graded to ensure that there are sufficient smaller stones to fill the voids between the larger ones

Coarse aggregates

Coarse aggregates are the larger particles of a concrete mix. They can be made from gravel or crushed rock. If gravel is to be used, this should also be crushed as the irregular shape of the particle gives a better bond with the cement than if it was left smooth. The size of the coarse aggregate can range between 5 and 40 mm but if it is to be used for reinforced concrete the size should be kept smaller than 20 mm.

Water

Water is used in the production of concrete to enable the cement to set and also to make the concrete 'workable'. Water must not contain any impurities, which might affect the strength of the concrete. The general rule for the quality of water is that it should be drinkable.

Mixes

Concrete is designed depending on where it is to be used. Mass concrete for normal strip foundations, oversite concrete, footpaths etc. should be mixed to a ratio of 1:3:6 of cement, fine aggregate and coarse aggregate. Concrete walls, beams and suspended floors etc. should have a ratio of 1:2:4 cement, fine aggregate and coarse aggregate.

For more structural concrete, the mixes are specified as a 'grade'. This grade is a number usually between 7 and 30 and is the amount of pressure given in newtons (N) that would be applied to each square millimetre of a cube of concrete before it is crushed in a test carried out after 28 days.

Batching

In order to produce concrete to a consistent strength and workability it is essential that each mix is accurately measured out or **batched** before being placed into the mixer.

Chapter 16 Basic concreting

Ready mix concrete can be delivered to site by lorry

The most accurate method of batching concrete is by weighing. On sites where large amounts of concrete are required each day, a mixer is used which can weigh the amount of aggregate placed in a hopper, before being fed into the mixer.

The ready mix concrete plants, which produce the concrete we see on the roads, in lorries and often on construction sites, would use a similar method, but on a larger scale.

Where weight batching is impossible, the next most accurate method is by volume. When volume batching, the amount of cement in a 25 kg bag would be 0.0175 m³, and so if we needed a 1:3:6 mix we would require three times this volume of fine aggregate and six times this volume of coarse aggregate.

To ensure accurate proportions of the aggregates a **gauge box** should be made for each of the aggregates.

 **Definition**

Gauge box – a bottomless box which should hold the correct volume of aggregate needed to mix a 25 kg bag of cement

Brickwork NVQ and Technical Certificate Level 2

The volume of the gauge box would be made to the ratios shown in Table 16.1 below.

Mix	Cement	Fine aggregate	Coarse aggregate
1:3:6	1 bag (25 kg)	0.050 m³	0.100 m³
1:2:4	1 bag (25 kg)	0.035 m³	0.070 m³

Table 16.1 Ratios for volume of gauge box

Did you know?

Many concreting jobs require constant mixes for strength and colour to be maintained, and so 'shovel fulls' are not an accurate enough method of batching. This is because a shovel full to one person may be more or less than a shovel full to another person who may mix the next batch. This would result in different proportions of mixes

When batching the concrete, the boxes are placed on a clean, firm base, filled with the aggregate and levelled off at the top. The area is cleaned of all spillages and the boxes are lifted off. The aggregates are then loaded into the mixer along with a full bag of cement.

Water-cement ratio

Water is used when mixing the concrete to cause the cement to set and make the concrete 'workable'. The more water in the mix, the more workable the concrete will be. However, the cement will only require a certain amount of water to make it set.

Any excess water is called **free water**. This free water occupies space and when the concrete dries out voids are left behind, causing the concrete to become weak.

The amount of water that should be added to a mix is given as a ratio between the cement and water, and this is called the **water-cement ratio**. This water-cement ratio is specified as a decimal number, usually between 0.4 and 0.8.

In order to work out the amount of water needed in a concrete mix, the water-cement ratio is multiplied by the weight of cement being mixed. This gives the amount of water in litres.

Chapter 16 Basic concreting

For example, if a concrete mix has a water cement ratio of 0.5 and contains 25 kg of cement, the amount of water needed for this mix would be:

0.5 x 25 = 12.5 litres

This is the total quantity of water that would be specified by an engineer.

Remember

Allowances must be made for the quantity of water present in the aggregates, i.e. damp sand

On the job: Mixing cement

Jimmy is mixing concrete on a building site. The concrete is for a new footpath around the perimeter of one of the new houses. After mixing for approximately 25 minutes, he realises that there are only six bags of cement left to use. He explains this to his boss who is laying the concrete. His boss tells him to use less per mix to make it go further. What should he do?

Types of concrete

Before starting a concreting project, it must be decided whether to use ready mix (mixed at a concrete plant and transported to site) or to self mix concrete. Each has advantages over the other.

Advantages of self mix:

1. cost is much lower
2. no need to order in advance (if materials are on site)
3. project can be postponed without advance notice (e.g. owing to bad weather).

Advantages of ready mix:

1. no need for storage of materials
2. able to provide large batches of materials
3. no need to employ labourers to mix the concrete.

Remember

For ready mix to be an advantage, the concrete lorry must have good access to the pour area

287

Formwork

The purpose of **formwork** is to hold the freshly placed and compacted concrete until it has set. To achieve this, the formwork should:

- be rigid enough to prevent bending
- strong enough to carry the weight of the concrete
- set in place to line and level
- have tight joints to prevent water or cement paste loss
- have suitable size panels to allow safe and easy handling
- be designed in a way that air pockets are not trapped.

Formwork may be made from timber or steel, and the choice of which to use usually depends on how many times the formwork is to be used. Steel formwork may be initially more expensive than timber but it can be reused more times as timber can be easily damaged during the striking of the formwork.

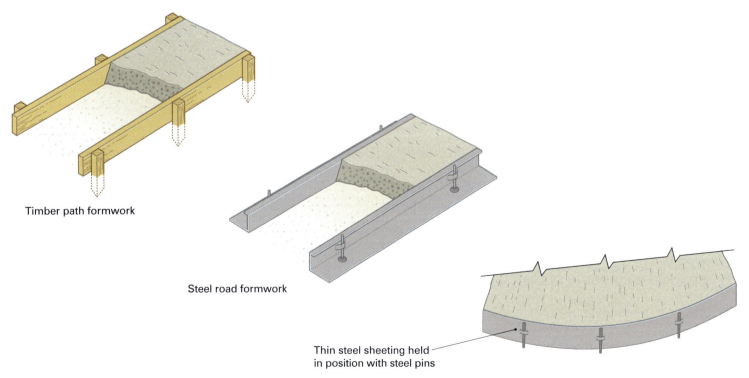

Figure 16.2 Types of formwork

Formwork for pathways

This formwork uses timber or steel road forms, with the height of the formwork being the same as the thickness of the concrete path. The formwork is positioned to align with the required amount of fall to allow for surface water drainage.

Pegs or steel pins are driven behind the formwork to prevent them from moving outwards when concrete is placed. Wedges may be used to adjust the alignment of the formwork.

Formwork for ground floors

Floors for buildings such as factories and warehouses etc. have large areas and would be difficult to lay in one slab. Floors of this type are usually laid in alternative strips up to 4.5 m wide, running the full length of the building (see Figure 16.3). The actual formwork would be similar to that used for paths.

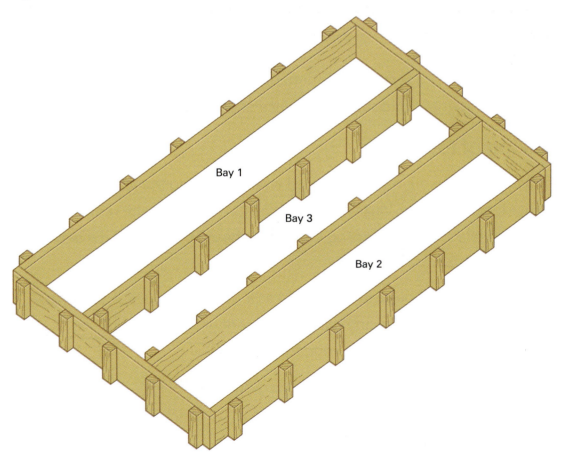

Figure 16.3 Alternative strip method used for large floor areas

Fixing formwork

The fixing of formwork may be made using nails, clamps or bolts. Clamps or bolts are preferred over nails as they are easier to strip off the formwork after completion, with less chance of anyone stepping on a nail which has been left sticking out of an old piece of formwork.

Release agent

To prevent damage to the concrete and allow for easier striking of the formwork, the surface of the formwork must be coated with a release agent before concrete is placed. Suitable release agents are:

- light oil
- chemical release agent
- mould cream emulsion
- water-based emulsion
- wax.

The type of release agent is dependent on the surface finish required as some of the agents could stain the surface of the concrete.

Striking of formwork

Striking the formwork is the term used when removing the formwork from the hardened concrete. Vertical sides of the formwork may be removed after 12 hours while soffits supporting lintels etc. should be left in place between 7 and 14 days.

All formwork should be cleaned as soon as it has been struck. A stiff brush should be used to remove any dust and cement grout. Stubborn bits of grout can be removed using a wooden scraper.

Did you know?

Another reason for not using nails is that, due to the amount of pressure on the formwork during compaction of the concrete, the joints could open, causing the concrete to leak

Remember

Steel tools should not be used to clean formwork as they can easily scratch the forms, which could show on the concrete surface

Safety tip

Use suitable PPE (personal protective equipment), especially eye and respiratory protection when using release agents and gloves, eye protection and a face mask when striking

Chapter 16 Basic concreting

Reinforcement

Concrete is strong in **compression** but weak in **tension** so, to prevent concrete from being 'pulled' apart when under pressure, steel reinforcement is provided. The type and position of the reinforcement will be specified by the structural engineer.

> **Definition**
>
> **Compression** – squeezed or squashed together
>
> **Tension** – stretched

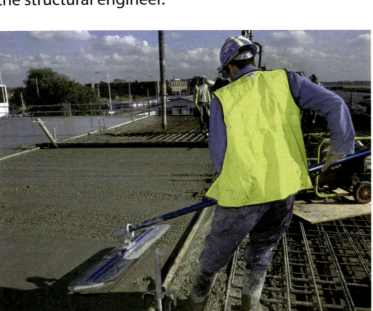

Steel reinforcement of concrete

The reinforcement must always have a suitable thickness of concrete cover to prevent the steel from rusting if exposed to moisture or air. The amount of cover required depends upon the location of the site with respect to exposure conditions, and ranges from 20 mm in mild exposure to 60 mm for very severe exposure to water.

To prevent the reinforcement from touching the formwork, spacers should be used. Made from concrete, fibre cement or plastic they are available in several shapes and various sizes to give the correct cover.

An example of concrete spacers

291

Expansion joints

Concrete expands and contracts as air temperatures rise and fall, so provision must be made for this variation in the length of the slab. These provisions are termed expansion joints and should be placed between 7 and 10 m apart. An expansion joint provides a gap between adjoining slabs, allowing each slab as it expands to take up this space and so prevent the concrete from cracking.

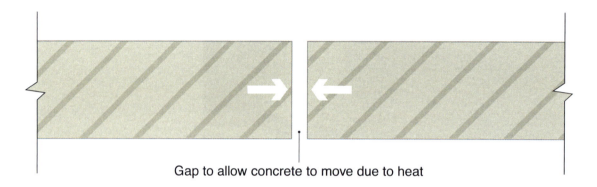

Figure 16.4 Concrete slabs without joint

The expansion joint is formed using a 12 mm thick fibreboard, 25 mm below the surface. The top of the expansion joint is filled with a flexible sealant. These materials are most suitable as they can be compressed.

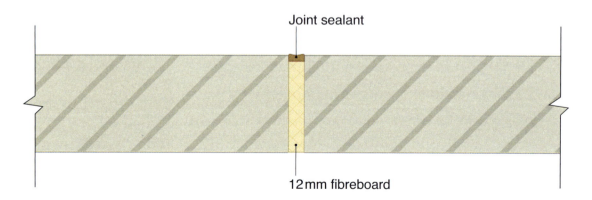

Figure 16.5 Concrete slabs with expansion joint

Workability

Mixed concrete must be sufficiently **workable** so that it can be fully compacted. Workability is affected by:

1. the water-cement ratio
2. the cement-aggregate ratio
3. the size of the coarse aggregate
4. the shape of the aggregate.

Workability may be measured by the slump test.

Definition

Workable – easy to use

Slump test

A sample of a batch of concrete is taken and placed in a steel cone 100 mm diameter at the top, 200 mm diameter at the bottom and 300 mm high, placed on a levelled base plate.

Step 1 Standing on the foot pieces, fill the cone in four equal layers.

Step 1 Filling the cone

Step 2 Rod each layer 25 times with a 16 mm diameter, rounded rod.

Step 2 Rodding

Step 3 Smooth off the concrete when the cone is full.

Step 3 Smoothing off

Step 4 Slowly lift the cone straight up and off.

Step 4 Lifting off the cone

Step 5 Lay the rod across the upturned slump cone and measure the distance between the concrete and the rod. This distance is the slump.

Step 5 Measuring the slump

There are three kinds of slump:

1. Normal
2. Collapse
3. Shear.

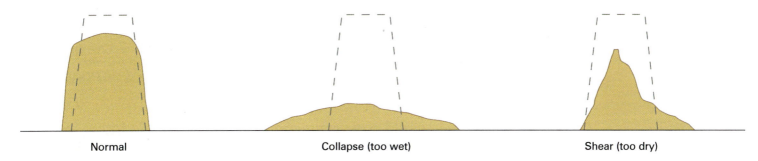

Figure 16.6 Slump test results

A concrete mix would have been designed to have a certain degree of slump. This would be given as the amount of workability. Table 16.2 shows the degrees of slump for concrete mixes.

Workability	Slump	
Low	10 – 30 mm	If the slump does not meet the specified slump, it should not be used.
Medium	25 – 75 mm	
High	65 –135 mm	

Table 16.2 Degrees of slump for concrete mixes

Test cubes

A test cube is a small metal box filled with a sample of concrete and used to the compression strength of concrete.

1. Samples of concrete should be taken and placed in 150 mm cube moulds. The concrete should be placed in the moulds in three layers, compacting each layer 25 times.
2. The cubes should be marked for identification while still wet.
3. The cubes are covered with a damp cloth and allowed to set.
4. When set, the mould is removed and the cubes are cleaned of surplus concrete and stored in a tank of warm water.

5. After 28 days the cubes are tested by a technician in a laboratory, where the cubes are put under pressure until they crush, to ensure the quality of the mix. If the cubes do not meet the compression force designed by the engineer, this could result in the whole structure having to be demolished and could end with a concrete supplier being found liable for the cost.

Compacting

When concrete has been placed, it contains trapped air in the form of voids. To get rid of these voids we must compact the concrete. The more workable the concrete the easier it would be to compact, but also if the concrete is too wet, the excess water will reduce the strength of the concrete.

Failure to compact concrete results in:

1. reduction in the strength of the concrete
2. water entering the concrete, which could damage the reinforcement
3. visual defects, such as honeycombing on the surface.

The method of compaction depends on the thickness and the purpose of the concrete. For oversite concrete, floors and pathways up to 100 mm thick, manual compaction with a tamper board may be sufficient. This requires slightly overfilling the formwork and tamping down with the tamper board. For larger spans the tamper board may be fitted with handles.

Tamper board with handles

For slabs up to 150 mm thick, a vibrating beam tamper should be used. This is simply a tamper board with a petrol-driven vibrating unit bolted on. The beam is laid on the concrete with its motor running and is pulled along the slab.

For deeper structures, such as retaining walls for example, a poker vibrator would be required. The poker vibrator is a vibrating tube at the end of a flexible drive connected to a petrol motor. The pokers are available in various diameters from 25 mm to 75 mm.

The concrete should be laid in layers of 600 mm with the poker in vertically and penetrating the layer below by 100 mm. The concrete is vibrated until the air bubbles stop and the poker is then lifted slowly and placed between 150 and 1000 mm from this incision, depending on the diameter of the vibrator.

Vibrating beam tamper

Vibrating poker in use

Surface finishes

Surface finishes for slabs may be:

1. Tamped finish. Simply using a straight edge or tamper board when compacting the concrete will leave a rough finish to the floor ideal for a path or drive surface, giving grip to vehicles and pedestrians. This finish may also be used if a further layer is to be applied to give a good bond.

2. Float and brush finish. After **screeding** off the concrete with a straight edge, the surface is floated off using a steel or wooden float and then brushed lightly with a soft brush (see photo below). Again, this would be suitable for pathways and drives.

3. Steel float finish. After screeding off using a straight edge, a steel float is applied to the surface. This finish attracts particles of cement to the surface, causing the concrete to become impermeable to water but also very slippery when wet. This is not very suitable for outside but ideal for use indoors for floors etc.

4. Power trowelling/float. Three hours after laying, a power float is applied to the surface of the concrete. After a further delay to allow surface water to evaporate, a power trowel is then used. A power float has a rotating circular disc or four large flat blades powered by a petrol engine. The edges of the blades are turned up to prevent them digging into the concrete slab. This finish would most likely be used in factories where a large floor area would be needed.

Definition

Screeding – levelling off concrete

Remember

Make sure you always clean all tampers and tools after use

Did you know?

The success of surface finishes depends largely on timing. You need to be aware of the setting times in order to apply the finish

Brushed concrete finish

5. Power grinding. This is a technique used to provide a durable wearing surface without further treatment. The concrete is laid, compacted and trowel finished. After 1 to 7 days the floor is ground, removing the top 1–2 mm, leaving a polished concrete surface.

Surface treatment for other surfaces may be:

1. Plain smooth surfaces. After the formwork has been struck, the concrete may be polished with a carborundum stone, giving a polished water-resistant finish.

2. Textured and profiled finish. A simple textured finish may be made by using rough sawn boards to make the formwork. When struck, the concrete takes on the texture of these boards. A profiled finish can be made by using a lining inside the formwork. The linings may be made from polystyrene or flexible rubber-like plastics, and gives a pattern to the finished concrete.

3. Ribbed finishes. These are made by fixing timber battens to the formwork.

4. Exposed aggregate finish. The coarse aggregate is exposed by removing the sand and cement from the finished concrete with a sand blaster. Another method of producing this finish is by applying a chemical retarder to the formwork, which prevents the cement in contact with it from hardening. When the formwork is removed, the mortar is brushed away to uncover the aggregate in the hardened concrete.

Power float

Ribbed concrete finish

Curing

Remember

If the water is allowed to evaporate from the mix shortly after the concrete is placed, there is less time for the cement to 'go off'

When concrete is mixed, the quantity of water is accurately added to allow for hydration to take place. The longer we can keep this chemical reaction going, the stronger the concrete will become.

To allow the concrete to achieve its maximum strength, the chemical reaction must be allowed to keep going for as long as possible. To do this we must 'cure' the concrete. This is done by keeping the concrete damp and preventing it from drying out too quickly.

Curing can be done by:

1. Spraying the concrete with a chemical sealer, which dries to leave a film of resin to seal the surface and reduces the loss of moisture.

2. Spraying the concrete with water, which replaces any lost water and keeps the concrete damp. This can also be done by placing sand or hessian cloth or other similar material on the concrete and dampening.

3. Covering the concrete with a plastic sheet or building paper, preventing wind and sun from evaporating the water into the air. Any evaporated moisture due to the heat will condense on the polythene and drip back on to the concrete surface.

Chapter 16 Basic concreting

Concreting in hot weather

When concreting in temperatures over 20°C, there is a reduction in workability due to the water being lost through evaporation. The cement also tends to react more quickly with water, causing the concrete to set rapidly.

To remedy the problem of the concrete setting quickly, a 'retarding mixture' may be used. This slows down the initial reaction between the cement and water, allowing the concrete to remain workable for longer.

Extra water may be added at the time of mixing so that the workability would be correct at the time of placing.

Water must *not* be added during the placing of the concrete, to make it more workable, after the initial set has taken place in the concrete.

Concreting in cold weather

Water expands when freezing. This can cause permanent damage if the concrete is allowed to freeze when freshly laid or in hardened concrete that has not reached enough strength (5 N/mm^2, which takes 48 hours).

Concreting should not take place when the temperature is 2°C or less. If the temperature is only slightly above 2°C, mixing water should be heated.

After being laid, the concrete should be kept warm by covering with insulating quilts, which allows the cement to continue its reaction with the water and prevents it from freezing.

The Cement & Concrete Association

The Cement & Concrete Association (C&CA) offers a service of technical information and advice, based on the work at its Research Station, combined with wide practical experience and the collection of information on a worldwide basis. The Association's training centre provides an extensive range of courses on concrete practice. Information is passed through publications and films.

Remember

In hot weather the concrete must be placed quickly and not left standing for too long

Safety tip

Take precautions in hot weather against the effect of the sun on your skin, for example wear sun block and a T-shirt

Find out

What kind of information could you get from C&CA? If you have access to the internet, you might like to visit their website. A link has been made available at www.heinemann.co.uk/hotlinks - just enter the express code 0866P.

FAQ

Are there special types of concrete available?

Special additives can be added to a concrete mix to give it particular properties, for example an anti-washout additive can be added to concrete that is to be under water and concrete can also be made to withstand very high temperatures if, for example, it is going to be used in chimney construction.

Knowledge check

1. Who enforces the Health and Safety at Work Act?
2. What is coarse aggregate made of?
3. What are the best methods of fixing formwork?
4. What is carried out to improve workability of concrete?
5. Why is reinforcement used in concrete?
6. What causes hydration?
7. Why is concrete compacted?
8. How long after cubes are made are they tested for strength?
9. What is the most common cement used in concrete?
10. How can you produce concrete to a consistent strength?
11. What is formwork used for?
12. What is used to allow variation of movement in large concrete slabs?
13. What are the three ingredients that make up concrete?
14. What is release agent used for?
15. What is carried out to check the cleanliness of aggregates?

Glossary

adhesive	glue
air bricks	bricks with holes to allow air to pass through a wall
arch centre	the temporary timber support for the arch ring on semi-circular arches during construction
architrave	a decorative moulding fitted around door and window frames
arris	the edge of a brick
autoclave	a machine that uses high pressure and temperature to cause chemical reactions
banding	sections of brickwork of slightly different shades
barrier cream	a cream used to protect the skin from damage or infection
batching	accurately measuring out concrete mixes by weight or volume
boundary lines	lines that dictate ownership of property
bridge	where moisture can be transferred from an outer wall to the inner leaf by material touching both walls
bucket handle finish	a shallow, round, inward joint that used to be created with a bucket handle
building line	an imaginary line set by the local authority to control the positioning of buildings
carded scaffolder	someone who holds a certificate showing competence in scaffold erection
cavity battens	a timber piece laid in a cavity to prevent mortar droppings falling down the cavity
cavity fill	concrete put into a cavity up to ground level
cavity liner	placed behind an air brick to form an air duct
compression	squeezed or squashed together
conservation	preservation of the environment and wildlife

contamination	the pollution of something
corbel	several course of bricks laid in front or behind the normal face line in order to produce a feature on a gable end
corrosive	a substance that damages things it comes into contact with
coving	a decorative moulding fitted at the top of a wall
crawling board	a board or platform used to support a worker's weight during roof work
damp proof course	a substance that is used to prevent damp from penetrating a building
dead men	temporary brick supports built on each abutment that provide an accurate straight edge
dermatitis	a skin condition characterised by red, dry and itchy skin
drip	see **throating**
dummy frame	a temporary frame put in place to ensure the opening dimensions and square are maintained during arch construction
egress	an exit
employer	the person or company you work for
enforced	making sure a law is obeyed
footings	brickwork between the foundation concrete and the horizontal DPC
formwork	timber or steel casing used to contain concrete while it hardens
foundations	concrete bases supporting walls
free water	excess water in a concrete mix
frontage line	the line of the front of a building
gauge box	a special box used to accurately measure materials
hazard	a danger or risk
header	the end of a brick

303

Health and Safety Executive (HSE)	the government organisation that enforces health and safety law in the UK
horizontal DPC	a layer of impervious material built into a wall 150 mm minimum above ground level
hydration	the addition of water to cement paste, which creates a chemical reaction that sets the mortar
in-situ concrete	concrete placed in its finished position
kinetic lifting	a way of lifting that reduces the risk of injury to the lifter
LPG	liquefied petroleum gas
main drain	a local authority main sewer
muster points	fire assembly points
noxious	harmful or poisonous
parapet	a low wall that acts as a barrier
perp	the vertical joint between two bricks or blocks
pier	a vertical support structure that gives strength to a wall
plant	industrial machinery
plaster skim	a thin layer of plaster
proactive	taking action before something happens
profile	a support for a line outside the working area
proportionately	in proportion to the size of something else
prosecute	to accuse someone of committing a crime
quoin	the corner of a wall
reactive	taking action after something happens
screeding	levelling off concrete or plaster
skirting	a decorative moulding fitted at the bottom of a wall
stiles	the side pieces of a ladder into which the steps are set
stretcher	the length of a brick
subsidence	the sinking in of buildings
symptom	a sign of illness or disease
Temporary Bench Mark (TBM)	a fixed levelling point to which other levelling points are related. Used on larger sites
tension	stretched
throating	(also known as drip) a groove under a window sill that prevents water from running across the bottom of the sill
tie-rods	metal rods underneath the rungs of a ladder
tingle plate	a flat piece of steel used to stop a string sagging
toxic	poisonous
trigonometry	the part of mathematics concerned with triangles and angles
true	when a measuring instrument or tool gives accurate readings
trusses	a roof component that spreads the load of the roof over the outer walls
turning piece	the temporary timber support for the arch ring on segmental arches during construction
vertebrae	the small bones that form the backbone
vertical DPC	a layer of impervious material placed up to the sides of openings in cavity walls to prevent the passage of moisture
vibration white finger	a painful condition that can be caused by using vibrating machinery
vitrified	a material that has been converted into a glass-like substance
voussoirs	wedge-shaped bricks used in axed arch construction
wall ties	metal or plastic fixings to tie cavity walls together
water-cement ratio	the amount of water added to a cement mix
weep holes	vertical joints left out in brickwork for water to run out
window head	the top of a window
workable	concrete or mortar that is easy to use

Index

3: 4: 5 method 150–1

A

abbreviations on drawings 94–5
accidents 41–2, 62–3
aggregates 126–7, 280, 281–4
angle grinders 116
apprenticeships 24–6
arches
    complex 244–5
    constructing simple 236–41
    geometry 232–5
    parts of 230–1
    temporary support 242–3
assembly drawings 88

B

batching concrete 284–6
bench marks 96–7, 144–5
block cutters 120
block plans 86
block work 177, 188–9
body language 16
bonding 170–6, 180, 193–4
breakers 117–18
brick cleaner 139
bricks 130–2, 171
    bonding 170–6
    laying 180–7
builder's square 149–50
building process 181–3
building regulations 11
building team 6–9
buildings, parts of/types of 10–12

C

cavity walls
    closing at eaves level 224–6
    constructing 208–16
    gable ends 226–8
    insulation types 217–19
    keeping clean 215–16
    openings in 219–28
cement 123, 128, 129, 161–2, 281
Cement and Concrete Association
    (C&CA) 301
chisels 110–11
CITB (Construction Industry Training
    Board) 27
colouring agents, mortar 163
communication 13–23
component range drawings 87–8
concrete 127, 280–4, 287

concreting
    compacting 296–7
    curing 300–1
    expansion joints 292
    formwork 288–90
    mixes 284–7
    reinforcement 291
    surface finishes 298–9
    workability 293–6
construction work, types of 2–3
contracts of employment 29–30
control of noise regulations 39
corner blocks 114, 185
corner profiles 115, 212–13
COSHH (Control of Substances Hazardous
    to Health) regulations 37–8, 280
crawling boards 75
curing concrete 300–1

D

damp penetration 220–4
damp proof course (DPC) 136–7, 209–10,
    222–4
datum points 95–7
day work sheets 20
dead men 228, 240
delivery notes 22, 124–5
discrimination at work 30
documents 18–23
drainage systems 256–9
    constructing 277
    excavating 268–9
    gradients 264–7
    inspection chambers 270–3
    pipes 134–5, 260–3, 274–5
    testing 276
    ventilation 259
drawings 18, 85
    abbreviations/symbols 93–5
    datum points 95–7
    equipment for 89–91
    scales 92–3
    title panels 89
    types 86–8
drills 119
dumpy levels 147–8

E

electricity at work regulations 40
employment 4–6
    contracts 29–30
    legislation 28–9
English bonds 173, 174

equipment 103
    for drawings 89–91
    hand tools 104–15
    portable power tools 116–20
eye protection 58–9

F

finishes, solid walls 201–4
fires, dealing with 50–3
Flemish bond 175
floors 210–11
foot protection 59
formwork 288–90
frames, door/window 219–20

G

gable walls 198–9, 225–8
gauge boxes 285–6
geometry 232–5

H

half bond 172
hammers 108–9
hand hawks 106
hand protection 61
hand tools 104–15
hardcore 128
head protection 57
health and safety 33, 248
    fires, handling 50–3
    handwashing 43
    inspectors 7
    laying bricks 180
    legislation 34–41, 280
    risks to workers 41–3
    setting out 142
    tools 104
    working at height 68–9
hearing protection 60
hop-ups 78
HSE (Health and Safety Executive) 35
hydration 162, 281

I

ill health 43
information sources 31
injuries, avoiding 45–9
insulation, types of 217–19
isometric projection 99–100

J

job sheets 20
jobs *see* employment
jointing 181, 248–9
    tools for 106
    types of finish 253–6

K

kerbs 134
kinetic lifting 47–9, 122, 180

L

ladders 71–4
laser levels 148
legislation
    employment 28–9
    health and safety 34–41
levelling compounds 130
lifting techniques 45–9, 122–3
lines and pins 113–14, 183–5
lintels 134, 223–4
liquid materials 138–9
location drawings 86–7

M

manual handling 39, 45–9, 122–4
materials
    delivery 124–5
    moving 122–4
    setting out 152
    storage 126–39
minimum wage 29
mixes
    concrete 284–7
    mortar 163–8, 250
mortar 126
    constituents 160–3
    gauging materials 165
    mixes/mixing 163–8, 250–1
    pre-mixed 167–8
moving materials 122–4

N

near misses, reporting 63
noise legislation 39
NVQs (National Vocational Qualifications) 26–7

O

operatives 6, 9
optical levels 146, 147
order forms 21
orthographic projection 98–9

P

paving slabs 133
petrol cutters 117
piers 195–6, 204–5
pipes see drainage systems
plaster 129
plasterboard 137–8

plasticisers 138, 163
platforms 76–7
plywood 137–8
pointing 249–50
portable power tools 116–20
PPE (Personal Protective Equipment) 40, 57–61
pre-cast concrete lintels 134
profiles 155, 156, 157
projections 98–100
PUWER (Provision and Use of Work Equipment Regulations) 38–9

Q

qualifications 26–8
quarter bond 173, 176

R

raking cutting 198–9, 227–8
raking out 250
reinforcement 197–8, 291
repointing 251–2
requisition forms 21
respiratory protection 61
RIDDOR, accident reporting law 41, 62
risk assessments 64–5, 68
roof work 75, 224–8
roofing tiles 136
rough ringed arches 237–40

S

safety signs 54–6
sand 126–7, 160
scabblers 119
scaffolding 78–82
scales on drawings 92–3
schedules 18
setting out 141
    health and safety 142
    keeping level 144–8
    keeping square 149–51
    materials needed 152
    positioning 143
    segmental arches 233–5
    step-by-step guide 153–7
silt test 160–1, 282–3
site datum 96–7, 144–5
site plans 87
site square 149
slump test 293–5
solid walls 191–2
    joining 199–200
    raking cutting 198–9

reinforcing 197–8
    types of 192–6
    vertical movement joints 201
    weathering 201–4
specifications 18, 100–1
spirit levels 112–13
stepladders 69–71
storage
    aggregates 127
    blocks/bricks 132–3
    cement and plaster 129
    drainage pipes 135, 261–2
    paving slabs 133
    roofing tiles 136
symbols on drawings 93–4

T

tape measures 107
technical certificates 27–8
timesheets 19
tingle plates 115, 186
title panels, drawings 89
tools
    hand tools 104–15
    power tools 116–20
tower scaffolds 80, 82
training 24–8
trestle platforms 76–7
trowels 105

V

verbal communication 13–14

W

wall ties 212, 213–15
walls
    cavity 207–28
    preventing movement 201
    protecting newly laid 187
    solid 191–205
water 162, 284
water-cement ratio 286–7
weather conditions
    protecting newly laid brickwork 187
    workability of concrete 301
weathering a wall 201–4
welfare facilities 44–5
work programmes 23
working at height 39–40, 67, 68–9
written communication 14–15